Functional Estimation for Density, Regression Models and Processes

Functional Estimation for Density, Regression Models and Processes

Odile Pons
INRA, France

NEW JERSEY • LONDON • SINGAPORE • BEIJING • SHANGHAI • HONG KONG • TAIPEI • CHENNAI

Published by

World Scientific Publishing Co. Pte. Ltd.
5 Toh Tuck Link, Singapore 596224
USA office: 27 Warren Street, Suite 401-402, Hackensack, NJ 07601
UK office: 57 Shelton Street, Covent Garden, London WC2H 9HE

British Library Cataloguing-in-Publication Data
A catalogue record for this book is available from the British Library.

FUNCTIONAL ESTIMATION FOR DENSITY, REGRESSION MODELS AND PROCESSES

ISBN-13 978-981-4343-73-2
ISBN-10 981-4343-73-0

Printed in Singapore by World Scientific Printers.

Preface

Nonparametric estimators have been intensively used for the statistical analysis of independent or dependent sequences of random variables and for samples of continuous or discrete processes. The optimization of the procedures is based on the choice of a bandwidth that minimizes an estimation error for functionals of their probability distributions.

This book presents new mathematical results about statistical methods for the density and regression functions, widely presented in the mathematical literature. There is no doubt that its origin benefits from earlier publications and from other subjects I worked about in other models for processes. Some questions of great interest for optimizing the methods have motivated much work some years ago, they are mentioned in the introduction and they give rise to new developments of this book. The methods are generalized to estimators with kernel sequences varying on the sample space and to adaptative procedures for estimating the optimal local bandwidth of each model.

More complex models are defined by several nonparametric functions or by vector parameters and nonparametric functions, such as the models for the intensity of point processes and the single-index regression models. New estimators are defined and their convergence rates are compared.

Odile M.-T. Pons

Contents

Chapter 1

Introduction

The aim of this book is to present in the same approach estimators for functions defining probability models: density, intensity of point processes, regression curves and diffusion processes. The observations may be continuous for processes or discretized for samples of densites, regressions and time series, with sequential observations over time. The regular sampling scheme of the time series is not common in regression models where stochastic explanatory variables X are recorded together with a response variable Y according to a random sampling of independent and identically distributed observations $(X_i, Y_i)_{i \le n}$. The discretization of a continuous diffusion process yields a regression model and the approximation error can be made sufficiently small to extend the estimators of the regression model to the drift and variance functions of a diffusion process. The functions defining the probability models are not specified by parameters and they are estimated in functional spaces.

This chapter is a review of well known estimators for density and regression functions and a presentation of models for continuous or discrete processes where nonparametric estimators are defined.

On a probability space $(\Omega, \mathcal{A}, P)$, let X be a random variable with distribution function $F(x) = \Pr(X \le x)$ and Lebesgue density f, the derivative of F. The empirical distribution function and the histogram are the simplest estimators of a distribution function and a density, respectively. With a sample $(X_i)_{i \le n}$ of the variable X, the distribution function $F(x)$ is estimated by $\widehat{F}_n(x)$, the proportion of observations smaller than x, which converges uniformly to F in probability and almost surely if and only F is continuous. A histogram with bandwidth h_n consists in a partition of the range of the observations into disjoint subintervals of length h_n where the density is estimated by the proportion of observations X_i in each subintervals, divided by h_n. The bandwidth h_n tends to zero as n tends to infinity and nh_n^2 tends to infinity, thus the size of the partition tends to infinity with the sample size. For a variable X defined in a metric space $(\mathbb{X}, \mathcal{A}, \mu)$, the histogram is the local nonparametric estimator defined by a set of neighbourhoods $\mathcal{V}_h = \{V_h(x), x \in \mathbb{X}\}$, with $V_h(x) = \{s; d(x, s) \leq h\}$ for the metric d of $(\mathbb{X}, \mathcal{A}, \mu)$

$$\widehat{f}_{n,h}(x) = (n \int_{V_h(x)} dF_X)^{-1} \sum_{i=1}^{n} 1_{\{X_i \in V_h(x)\}}. \tag{1.1}$$

The empirical distribution function and the histogram are stepwise estimators and smooth estimators have been later defined for regular functions.

1.1 Estimation of a density

Several kinds of smooth methods have been developed. The first one was the projection of functions onto regular and orthonormal bases of functions $(\phi_k)_{k \geq 0}$. The density of the observations is approximated by a countable projection on the basis $f_n(x) = \sum_{i=1}^{K_n} a_k \phi_k(x)$ where K_n tends to infinity and the coefficients are defined by the scalar product specific to the orthonormality of the basis with

$$\int \phi_k^2(x) \mu_\phi(x)\, dx = 1, \int \phi_k(x) \phi_l(x) \mu_\phi(x)\, dx = 0, \text{ for all } k \neq l,$$

then $a_k =< f, \phi_k >= \int f(x) \phi_k(x) \mu_\phi(x)\, dx$. The coefficients are estimated by integrating the basis with respect to the empirical distribution of the variable X

$$\widehat{a}_{kn} = \int \phi_k(x) \mu_\phi(x)\, d\widehat{F}_n(x)$$

which yields an estimator of the density $\widehat{f}_n(x) = \sum_{i=1}^{K_n} \widehat{a}_{kn} \phi_k(x)$. The same principle applies to other stepwise estimators of functions. Well known bases of L_2-orthogonal functions are

(i) Legendre's polynomials[1] defined on the interval $[-1, 1]$ as solutions of the differential equations

$$(1 - x^2) P_n''(x) - 2x\, P_n'(x) - n(n+1) P_n(x) = 0,$$

[1] French mathematician (1752-1833)

with $P_n(1) = 1$. Their solutions have an integral form attributed to Hermite and his student Stieltjes

$$P_n(\cos\theta) = \frac{2}{\pi}\int_0^{\pi} \frac{\sin(n+\frac{1}{2})\phi\,d\phi}{\{2\cos\theta - 2\cos\phi\}}.$$

The polynom $P_n(x)$ has also been expressed as the coefficient of $z^{-(n+1)}$ in the expansion of $(z^2 - 2xz + 1)^{-1/2}$ by Stieltjes (1890). They are orthogonal with the scalar product

$$< f, g >= \int_{-1}^{1} f(x)g(x)\,dx;$$

(ii) Hermite's polynomials[2] of degree n defined by the derivatives

$$H_n(x) = (-1)^n e^{x^2/2}\frac{d^n}{dx^n}(e^{-x^2/2}),\ n \geq 1,$$

they satisfy the recurrence equation $H_{n+1}(x) = xH_n(x) - H_n'(x)$, with $H_0(x) = 1$. They are orthogonal with the scalar product

$$< f, g >= \int_{-\infty}^{+\infty} f(x)g(x)e^{-x^2}\,dx$$

and their norm is $\|H_n\| = n!\sqrt{2\pi}$;

(iii) Laguerre's polynomials[3] defined by the derivatives

$$L_n(x) = \frac{e^x}{n!}\frac{d^n}{dx^n}(e^{-x}x^n),\ n \geq 1,$$

and $L_0(x) = 1$. They satisfy the recurrence equation $L_{n+1}(x) = (2n + 1 - x)L_n(x) - n^2 L_{n-1}(x)$ and they are orthogonal with the scalar product

$$< f, g >= \int_{-\infty}^{+\infty} f(x)g(x)e^{-2x}\,dx.$$

The orthogonal polynomials are normalized by their norm. If the function f is Lipschitz, the polynomial approximations converge to f in L_2 and for the pointwise convergence. The corresponding projection estimators also converge in L_2 and pointwisely. Though the bases generate functional spaces of smooth integrable functions, the estimation is parametric. The estimator of the approximation function converge to zero in L_2 with the

[2]French mathematician (1822-1901)
[3]French mathematician (1834-1886)

norm $\|\widehat{f}_n - f_n\|_2 = \{\int_{-\infty}^{+\infty} E(\widehat{f}_n - f_n)^2(x)\mu_\phi(x)\,dx\}^{1/2}$ if $n^{-1}K_n$ tends to zero, so that

$$
\begin{aligned}
\|\widehat{f}_n - f_n\|_2^2 &= \int_{-\infty}^{+\infty} E\sum_{i=1}^{K_n}(\widehat{a}_{kn} - a_k)^2\phi_k^2(x)\mu_\phi(x)\,dx \\
&= \sum_{i=1}^{K_n} E(\widehat{a}_{kn} - a_k)^2 \\
E(\widehat{a}_{kn} - a_k)^2 &= E\{\int \phi_k(x)\mu_\phi(x)\,d(\widehat{F}_n - F)(x)\}^2 \\
&= n^{-1}\int \phi_k(x)\phi_k(y)\mu_\phi(x)\mu_\phi(y)\,dC(x,y)
\end{aligned}
$$

where $C(x,y) = F(x \wedge y) - F(x)F(y)$ is the covariance function of the empirical process $n^{1/2}(\widehat{F}_n - F)$. The convergence rate of the norm the density estimator is the sum of the norm $\|\widehat{f}_n - f_n\|_2 = O(n^{-1/2}K_n^{1/2})$ and the approximation error $\|f_n - f\|_2 = (\sum_{i=K_n+1}^{\infty} a_k^2)^{1/2}$, it is determined by the convergence rate of the sum of the squared coefficients and therefore by the degree of derivability of the function f. Splines are also bases of functions constrained at fixed points or by a condition of derivability of the function f, with an order of integration for its higher derivative. They have been introduced by Whittaker (1923) and developed by Schoenberg (1964), Wold (1975), Wahba and Wold (1975), De Boor (1978), Wahba (1978), Eubank (1988). They allow the approximation of functions having different degrees of smoothness on different intervals which can be fixed. A comparison between splines and kernel estimators of densities may be found in Silverman (1984) who established an uniform asymptotic bound for the difference between the kernel function and the weight function of cubic splines, with a bandwidth kernel $\lambda^{-1/4}$ where λ is the smoothing parameters of the splines. Each spline operator corresponds to a kernel operator and the bias and variance of both estimators have the same rate of convergence (Rice and Rosenblatt, 1983, Silverman, 1984). Messer (1991) provides an explicit expression of the kernel corresponding to a cubic sinusoïdal spline, with their rates of convergence.

Kernel estimators of densities have first been introduced and studied by Rosenblatt (1956), Whittle (1958), Parzen (1962), Watson and Laedbetter (1963), Bickel and Rosenblatt (1973). Consider a real random variable X defined on $(\Omega, \mathcal{A}, P)$ with density f_X and distribution function F_X. A continuous density f_X is estimated by smoothing the empirical distribution function $\widehat{F}_{X,n}$ of a sample $(X_i)_{1\leq i\leq n}$ distributed as X by the means of its convolution with a kernel K, over a bandwidth $h = h_n$ tending to zero as n tends to infinity

$$\widehat{f}_{X,n,h}(x) = \int K_h(x-s)\,d\widehat{F}_{X,n}(s) = \frac{1}{nh}\sum_{i=1}^{n} K\Big(\frac{x-X_i}{h}\Big), \tag{1.2}$$

where $K_h(x) = h^{-1}K(h^{-1}x)$ is the kernel of bandwidth h. The weighting kernel is a bounded symmetric density satisfying regularity properties and moment conditions. With a p-variate vector X, the kernel may be defined on $\mathbb{R}^p$ and $K_h(x) = (h_1 \ldots, h_p)^{-1}K(h_1^{-1}x_1, \ldots, h_p^{-1}x_p)$, for p-dimensional vectors $x = (x_1, \ldots, x_p)$ and $h = (h_1, \ldots, h_p)$. Scott (1992) gives a detailed presentation of the multivariate density estimators with graphical visualizations. Another estimator is based on the topology of the space $(\mathbb{X}, \mathcal{A}, \mu)$, with (1.1) or using a real function K and

$$K_h(x) = h^{-1}K(h^{-1}\|x\|_\mu),\ h > 0.$$

The regularity of the kernel K entails the continuity of the estimator $\widehat{f}_{X,n,h}$. All results established for a real valued variable X apply straightforwardly to a variable defined in a metric space.

Deheuvels (1977) presented a review of nonparametric methods of estimation for the density and compared the mean squared error of several kernel estimators including the classical polynomial kernels which do not satisfy the above conditions, some of them diverge and their orders differ from those of the density kernels. Classical kernels are the normal density with support $\mathbb{R}$ and densities with a compact support such as the Bartlett-Epanechnikov kernel with support $[-1, 1]$, $K(u) = 0.75(1-u^2)1_{\{|u|\leq 1\}}$, other kernels are presented in Parzen (1962), Prakasa Rao (1983), etc. With a sequence h_n converging to zero at a convenient rate, the estimator $\widehat{f}_{X,n,h}$ is biased, with an asymptotically negligible bias depending on the regularity properties of the density. Constants depending on moments of the kernel function also appear in the bias function $E\widehat{f}_{X,n,h} - f_X$ and the moments $E\widehat{f}^k_{X,n,h}$ of the estimated density. The variance does not depend on the class of the density. The weak and strong uniform consistency of the kernel density estimator and its derivatives were proved by Silverman (1978) under derivability conditions for the density. Their performances are measured by several error criteria corresponding to the estimation of the density at a single point or over its whole support. The mean squared error criterion is common for that purpose and it splits into a variance and the square of a bias term

$$\begin{aligned} MSE(\widehat{f}_{X,n,h};x,h) &= E\{\widehat{f}_{X,n,h}(x) - f_X(x)\}^2 \\ &= E\{\widehat{f}_{X,n,h}(x) - E\widehat{f}_{X,n,h}(x)\}^2 \\ &\quad + \{E\widehat{f}_{X,n,h}(x) - f_X(x)\}^2. \end{aligned}$$

A global random measure of the distance between the estimator $\widehat{f}_{X,n,h}$ and the density f_X is the integrated squared error (ISE) given by

$$ISE(\widehat{f}_{X,n,h};h) = \int \{\widehat{f}_{X,n,h}(x) - f_X(x)\}^2\,dx. \tag{1.3}$$

A global error criterion is the mean integrated squared error introduced by Rosenblatt (1956)

$$MISE(\widehat{f}_{X,n,h};h) = E\{ISE(\widehat{f}_{X,n,h};h)\} = \int MSE(\widehat{f}_{X,n,h};x,h)\,dx. \tag{1.4}$$

The first order approximations of the MSE and the MISE as the sample size increases are the AMSE and the AMISE. Let $(h_n)_n$ be a bandwidth sequence converging to zero and such that nh tends to infinity and let K be a kernel satisfying $m_{2K} = \int x^2K(x)\,dx < \infty$ and $\kappa_2 = \int K^2(x)\,dx < \infty$. Consider a variable X such that EX^2 is finite and the density F_X is twice continuously differentiable

$$AMSE(\widehat{f}_{X,n,h});x = (nh)^{-1}f_X(x)\kappa_2 + \frac{h^4}{4}m_{2K}^2f''^2(x).$$

They depend on the bandwidth h of the kernel and the AMSE is minimized at a value

$$h_{AMSE}(x) = \{f_X(x)\frac{\int K^2(x)\,dx}{nm_{2K}^2f''^2(x)}\}^{1/5}.$$

The global optimum of the AMISE is attained at

$$h_{AMISE} = \{\frac{\int K^2(x)\,dx}{nm_{2K}^2\int f''^2(x)\,dx}\}^{1/5}.$$

Then the optimal AMSE tends to zero with the order $n^{-4/5}$, it depends on the kernel and on the unknown values at x of the functions f_X and f''^2, or their integrals for the integrated error (Silverman, 1986). If the bandwidth has a smaller order, the variance of the estimator is predominant in the expression of the errors and the variations of estimator are larger, if the bandwidth is larger than the optimal value, the bias increases and the variance is reduced. The approximation made by suppressing the higher order terms in the expansions of the bias and the variance of the density estimator is obviously another source of error in the choice of the bandwidth, Hall and Marron (1987) proved that h_{MISE}/h_{AMISE} tends to 1 and h_{ISE}/h_{AMISE} tends to 1 in probability as n tends to infinity. Surveys on kernel density estimators and their risk functions were given by Nadaraya (1989), Rosenblatt (1956, 1971), Prakasa Rao (1983), Hall (1984), Härdle (1991), Khasminskii (1992). The smoothness conditions for the density are sometimes replaced by Lipschitz or Hölder conditions and the expansions for the MSE are replaced by expansions for an upper bound. Parzen (1962) also proved the weak convergence of the mode of a kernel density estimator. The derivatives of the density are naturally estimated by those of the kernel estimator and the weak and strong convergence of derivative estimators have been considered by Bhattacharya (1967) and Schuster (1969) among others. The $L_1(\mathbb{R})$ norm of the difference between the kernel estimator and its expectation converges to zero, as a consequence of the properties of the convolution. Devroye (1983) studied the consistency of the L_1-norm $\|\widehat{f}_{X,n,h} - f\|_1 = \int |\widehat{f}_{X,n,h} - f|\,dx$, Giné, Mason and Zaitsev (2003) established the weak convergence of the process $n^{1/2}(\|\widehat{f}_{X,n,h} - E\widehat{f}_{X,n,h}\|_1 - E\|\widehat{f}_{X,n,h} - E\widehat{f}_{X,n,h}\|_1)$ to a normal variable with variance depending on $\int K(u)K(u+t)\,du$. Bounds for minimax estimators have been established by Beran (1972). A minimax property of the kernel estimator with the optimal convergence rate $n^{2/5}$ was proved by Bretagnole and Huber (1981).

Though the estimator of a monotone function is monotone with probability tending to 1 as the number of observations tends to infinity, the number of observations is not always large enough to preserve this property and a monotone kernel estimator is built for monotone density functions by isotonisation of the classical kernel estimator. Monotone estimators for a distribution function and a density have been first defined by Grenander (1956) as the least concave minorant of the empirical distribution function and its derivative. This estimator has been studied by Barlow, Bartholomew, Bremmer and Brunk (1972), Kiefer and Wolfowitz (1976), Groeneboom (1989), Groeneboom and Wellner (1997). The isotonisation of the kernel estimator $\widehat{f}_{n,h}$ for a density function is

$$\widehat{f}_{SI,n,h}(x) = \inf_{v \geq x} \sup_{u \leq x} \frac{1}{v-u} \int_u^v \widehat{f}_{n,h}(t)\,dt \tag{1.5}$$

and $\int 1_{\{t \leq x\}} \widehat{f}_{SI,n,h}(t)\,dt$ is the greatest convex minorant of the integrated estimator $\int 1_{\{t \leq x\}} \widehat{f}_{n,h}(t)\,dt$. Its convergence rate is $n^{1/3}$ (van der Vaart and van der Laan, 2003). Groeneboom and Wellner studied the weak convergence of local increments of the isotonic estimator of the distribution function. The estimation of a convex decreasing and twice continuously differentiable density on $\mathbb{R}_+$ by a piecewise linear estimator with knots between observations points was studied by Groeneboom, Jonkbloed and Wellner (2001), the estimator is $n^{2/5}$-consistent. Dumbgen and Rufibach (2009) proposed a similar estimator for a log-concave density on $\mathbb{R}_+$ and established its convergence rate $(n(\log n)^{-1})^{\beta/(2\beta+1)}$, for Hölder densities of $\mathcal{H}_{\beta,M}$.

Stone (1974), De Boor (1975), Bowman (1983), Marron (1987) introduced automatic data driven methods for the choice of the global bandwidth. They minimize the integrated random risk ISE or the cross-validation criterion $CV(h) = \int \widehat{f}^2_{X,n,h}(x)\,dx - 2n^{-1}\sum_{i=1}^n \widehat{f}_{X,n,h,i}(X_i)$ where $\widehat{f}^2_{X,n,h,i}$ is the kernel estimator based on the data sample without the i-th observation, or the empirical version of the Kullback-Leibler loss-function $K(\widehat{f}_{X,n,h}, f) = -E\int \log \widehat{f}_{X,n,h}\,dF_X$ (Bowman, 1983). The $CV(h)$ criterion is an unbiased estimator of the MISE and its minimum is the minimum for the estimated ISE using the empirical distribution function. The global bandwidth estimator $\widehat{h}_{CV}$ minimizing this estimated criterion achieves the bound for the convergence rate of any optimal bandwidth for the ISE, $\widehat{h}_{CV}/h_{MISE} - 1 = O_p(n^{-1/10})$ and $\widehat{h}_{CV} - h_{MISE}$ has a normal asymptotic distribution (Hall and Marron, 1987). The cross-validation is more variable with the data and often leads to oversmoothing or undersmoothing (Hall and Marron, 1987, Hall and Johnstone, 1992). As noticed by Hall and Marron (1987), the estimation of the density and the mean squared error are different goals and the best bandwidth for the density might not be the optimal for the MSE, hence the bandwidth minimizing the cross-validation induces variability of the density estimator. Other methods for selecting the bandwidth have been proposed such as higher order kernel estimators of the density (Hall and Marron, 1987, 1990) or bootstrap estimations. An uniform weak convergence of the distribution of $\widehat{f}_{n,h}$ was proved using consecutive approximations of empirical processes by Bickel and Rosenblatt (1973), other approaches for the convergence in distribution rely on the small variations of moments of the sample-paths, as in Billingsley (1968) for continuous processes. The Hellinger distance $h(\widehat{f}_{X,n,h}, f)$ between a density and its estimator has been studied by Van de Geer (1993, 2000), here the weak convergence of the process $\widehat{f}_{X,n,h} - f_X$ provides a more precise convergence rate for $h(\widehat{f}_{X,n,h}, f)$. All result are extended to the limiting marginal density of a continuous process under ergodicity and mixing conditions.

Uniform strong consistency of the kernel density estimator requires stronger conditions, results can be found in Silverman (1978), Singh (1979), Prakaso Rao (1983), Härdle, Janssen and Serfling (1988) for the strong consistency and, for its conditional mode, Ould Saïd (1997). The law of the iterated logarithm has been studied by Hall (1981), Stute (1982), Bosq (1998)

Theorem 1.1 (Stute 1982). *Let f be a continuous density strictly positive and bounded on a sub-interval $[a,b]$ of its support. Let $(h_n)_n$ be a bandwidth sequence converging to zero and such that nh_n tends to infinity, $\log h_n^{-1} = o(nh_n)$ and $\log h_n^{-1}/(\log\log n)$ tends to infinity. Suppose that K has a compact support and $\int |dK| < \infty$, then for every $\delta > 0$*

$$\limsup_n \{\frac{nh_n}{2\log h_n^{-1}}\}^{1/2} \sup_{I_h} |\widehat{f}_{X,n,h}(x) - E\widehat{f}_{X,n,h}(x)| f^{1/2}(x) = \kappa_2^{1/2}, a.s.$$

with $I_h = [a+h, b-h]$.

A periodic density f on an interval $[-T,T]$ is analyzed in the frequency domain where it is expanded according to the amplitudes and the frequency or period of its components. Let $T = 2\pi/w$, the density f is expressed as the limit of series due to Fourier[4], $f(x) = \sum_{k=-\infty}^{+\infty} c_k e^{iwkx}$ with coefficients $c_k = T^{-1}\int_{-T/2}^{T/2} f(x)e^{-iwkx}\,dx$ and the Fourier transform of f is defined on $\mathbb{R}$ by $\mathcal{F}f(s) = T^{-1}\int_{-T/2}^{T/2} f(x)e^{-iwsx}\,dx$. The inversion formula of the Fourier transform is $f(x) = \int_{-\infty}^{+\infty} \mathcal{F}f(w)e^{iwsx}\,ds$. For a non periodic density, the Fourier transform and its inverse are defined by

$$\mathcal{F}f(s) = (2\pi)^{-1}\int_{-\infty}^{\infty} f(x)e^{-isx}\,dx,$$

$$f(x) = \int_{-\infty}^{+\infty} \mathcal{F}f(w)e^{isx}\,ds.$$

The Fourier transform is an isometry as expressed by the equality $\int |\mathcal{F}f(s)|^2\,ds = \int |f(s)|^2\,ds$.

Let $(X_k)_{k\le n}$ be a stationary time series with mean zero, the spectral density is defined from the autocorrelation coefficients $\gamma_k = E(X_0X_k)$ by $S(w) = \sum_{k=-\infty}^{+\infty} \gamma_k e^{-iwk}$ and the inverse relationship for the autocorrelations is $\gamma_k = \int_{-\infty}^{\infty} S(w)e^{iwsx}\,dx$. The periodogram of the series is defined as $\widehat{I}_n(w) = T^{-1}|\sum_{k=1}^n X_k e^{-2\pi ikw}|^2$ and it is smoothed to yield a regular estimator of the spectral density $\widehat{S}_n(s) = \int K_h(u-s)\widehat{I}_n(s)\,ds$. Brillinger (1975)

[4] French mathematician (1768–1830)

established that the optimal convergence rate for the bandwidth is $h_n = O(n^{-1/5})$ under regularity conditions and he proved the weak convergence of the process $n^{2/5}(\widehat{S}_n - S)$ to the process defined as a transformed Brownian motion. Robinson (1986, 1991) studied the consistency of kernel estimators for auto-regression and density functions and for nonparametric models of time series. Cross-validation for the choice of the bandwidth was also introduced by Wold (1975). For time series, Chiu (1991) proposed a stabilized bandwidth criterion having a relative convergence rate $n^{-1/2}$ instead of $n^{-1/10}$ for the cross-validation in density estimation. It is defined from the Fourier transform d_Y of the observation series $(Y_i)_i$, using the periodogram of the series $I_Y = d_Y^2/(2\pi n)$ and the Fourier transform $W_h(\lambda)$ of the kernel.The squared sum of errors is equal to $2\pi \sum_i I_Y(\lambda_j)\{1 - W_h(\lambda_j)\}^2$ with $\lambda_j = 2\pi j$ and $W_h(\lambda) = n^{-1} \sum_{j=1}^n \exp(-i\lambda j) K_h(j/n)$.

Multivariate kernel estimators are widely used in the analysis of spatial data, in the comparison and the classification of vectors.

1.2 Estimation of a regression curve

Consider a two-dimensional variable (X, Y) defined on $(\Omega, \mathcal{A}, P)$, with values in $\mathbb{R}^2$. Let f_X and $f_{X,Y}$ be the continuous densities of X and, respectively, f_{XY}, and let F_X and F_{XY} be their distribution functions. In the nonparametric regression setting, the curve of interest is the relationship between two variables, Y a response variable for a predictor X. A continuous curve is estimated by the means of a kernel estimator smoothing the observations of Y for observations of X in the neighborhood of the predictor value. The conditional mean of Y given $X = x$ is the nonparametric regression function defined for every x inside the support of X by

$$m(x) = E(Y|X = x) = \int y \frac{f_{X,Y}(x, y)}{f_X(x)}\, dy,$$

it is continuous when the density $f_{X,Y}$ is continuous with respect to its first component. It defines regression models for Y with fixed or varying noises according to the model for its variance

$$Y = m(X) + \sigma\varepsilon \tag{1.6}$$

where $E(\varepsilon|X) = 0$ and $Var(\varepsilon|X) = 1$ in a model with a constant variance $Var(Y|X) = \sigma^2$, or

$$Y = m(X) + \sigma(X)\varepsilon \tag{1.7}$$

with a varying conditional variance $Var(Y|X) = \sigma^2(X)$. The regression function of (1.6) is estimated by the integral of Y with respect to a smoothed empirical distribution function of Y given $X = x$

$$\widehat{m}_{n,h}(x) = \int y \frac{K_h(x-s)\widehat{F}_{XY,n}(ds,dy)}{\widehat{f}_{X,n,h}(x)}$$
$$= \frac{\sum_{i=1}^n Y_i K_h(x-X_i)}{\sum_{i=1}^n K_h(x-X_i)}.$$

This estimator has been introduced by Watson (1964) and Nadaraya (1964) and detailed presentations can bee found in the monographs by Eubank (1977), Nadaraya (1989) and Härdle (1990). The performance of the kernel estimator for the regression curve m is measured by error criteria corresponding to the estimation of the curve at a single point or over its whole support, like for the kernel estimator of a continuous density.

A global random measure of the distance between the estimator $\widehat{m}_{n,h}$ and the regression function m is the integrated squared error (ISE)

$$ISE(\widehat{m}_{n,h}; h) = \int \{\widehat{m}_{n,h}(x) - m(x)\}^2 \, dx, \tag{1.8}$$

its convergence was studied by Hall (1984), Härdle (1990). The mean squared error criterion develops as the sum of the variance and the squared bias of the estimator

$$MSE(\widehat{m}_{n,h}; x, h) = E\{\widehat{m}_{n,h}(x) - m(x)\}^2$$
$$= E\{\widehat{m}_{n,h}(x) - E\widehat{m}_{n,h}(x)\}^2 + \{E\widehat{m}_{n,h}(x) - m(x)\}^2.$$

A global mean squared error is the mean integrated squared error

$$MISE(\widehat{m}_{n,h}; h) = E\{ISE(\widehat{m}_{n,h}; h)\} = \int MSE(\widehat{m}_{n,h}; x, h) \, dx. \tag{1.9}$$

Assuming that the curve is twice continuously differentiable, the mean squared error is approximated by the asymptotic MSE (Chapter 3)

$$AMSE(\widehat{m}_{n,h}; x) = (nh)^{-1}\kappa_2 f_X^{-1}(x)\, Var(Y|X=x)$$
$$+ \frac{h^4}{4} m_{2K}^2 f_X^{-1}(x)\{\mu^{(2)}(x) - m(x) f_X^{(2)}(x)\}^2.$$

The AMSE is minimized at a value $h_{m,AMSE}$ which is still of order $n^{-1/5}$ and depends on the value at x of the functions defining the model and their second order derivatives. Automatic optimal bandwidth selection by cross-validation was developed by Härdle, Hall and Marron (1988) similarly to the density. Bootstrap methods were also widely studied. Splines were generalized to nonparametric regression by Wahba and Wold (1975), Silverman (1985) for cubic splines and the automatic choice of the degree of smoothing is also determined by cross-validation.

In model (1.7) with a random conditional variance $Var(Y|X) = \sigma^2(X)$, the estimator of the regression curve m has to be modified and it is defined as a weighted kernel estimator with weighting function $w(x) = \sigma^{-1}(x)$

$$\widehat{m}_{w,n,h}(x) = \frac{\sum_{i=1}^n w(X_i)Y_iK_h(x - X_i)}{\sum_{i=1}^n w(X_i)K_h(x - X_i)}$$

or more general function w. In Chapter 3, a kernel estimator of $\sigma^{-1}(x)$ is introduced. The bias and variance of the estimator $\widehat{m}_{w,n,h}$ are developed by the same expansions as the estimator (1.8). The convergence rate of the kernel estimator for $\sigma^2(x)$ is nonparametric and its bias depends on the bandwidths used in its definition, on $Var\{(Y - m(x))^2|X = x\}$, on the functions f_X, σ^2, m and their derivatives.

Results about the almost sure convergence and the L_2-errors of kernel estimators, their optimal convergence rates and the optimal bandwidth selection were introduced in Hall (1984), Nadaraya (1964). Properties similar to those of the density are developed here with sequences of bandwidths converging with specified rates. The methods for estimating a density and a regression curve by the means of kernel smoothing have been extensively presented in monographs by Nadaraya (1989), Härdle (1990, 1992) Wand and Jones (1995), Simonoff (1996), Bowman and Azalini (1997), among others. In this book, the properties of the estimators are extended with exact expansions, as for density desimation, and to variable bandwidth sequences $(h_n(x))_{n\geq 1}$ converging with a specified rate.

Several monotone kernel estimators for a regression function m have been considered, they are built by kernel smoothing after an isotonisation of the data sample, or by an isotonisation of the classical kernel estimator. The isotonisation of the data consists in a transformation of the observation $(Y_i)_i$ in a monotone set $(Y_i^*)_i$. It is defined by

$$Y_i^* = \min_{v\geq i} \max_{u\leq i} \frac{1}{v-u} \sum_{j=u}^{v} Y_i,$$

and $\sum_{i\leq k} Y_i^*$ is the greatest convex minorant of $\sum_{i\leq k} Y_i$. The kernel estimator for the regression function built with the isotonic sample $(X_i^*, Y_i^*)_i$ is denoted $\widehat{m}_{IS,n,h}$. The convergence rate of the isotonic estimator for a monotone density function is $n^{1/3}$ and the variable $n^{-1/3}(\widehat{m}_{IS,n,h} - m_{IS,n,h})(x)$

converges to a Gaussian process for every x in $\mathcal{I}_X$. The isotonisation of the kernel estimator $\widehat{m}_{n,h}$ for a regression function is

$$\widehat{m}_{SI,n,h}(x) = \inf_{v \geq x} \sup_{u \leq x} \frac{1}{v-u} \int_u^v \widehat{m}_{n,h}(t)\, dt \tag{1.10}$$

and $\int 1_{\{t \leq x\}} \widehat{m}_{SI,n,h}(t)\, dt$ is the greatest convex minorant of the process $\int 1_{\{t \leq x\}} \widehat{m}_{n,h}(t)\, dt$. Its convergence rate is again $n^{1/3}$ (van der Vaart and van der Laan, 2003). Meyer and Woodroof (2000) generalized the contraints to larger classes and proved that the variance of the maximum likelihood estimator of a monotone regression function attains the optimal convergence rate $n^{1/3}$.

In the regression models (1.6) or (1.7) with a multidimensional regression vector X, a multidimensional regression function $m(X)$ can be replaced by a semi-parametric single-index model $m(x) = g(\theta^T x)$, where θ^T denotes the transpose of a vector θ, or by a more general transformation model $g \circ \varphi_\theta(X)$ with unknown function m and parameter θ. In the single-index model, several estimators for the regression function $m(x)$ have been defined (Ihimura, 1993, Härdle, Hall and Ihimura, 1993, Hristache, Juditski and Spokony, 2001, Delecroix, Härdle and Hristache, 2003), the estimators of the function g and the parameter θ are iteratively calculated from approximations.

The inverse of the distribution function F_X of a variable X, or quantile function, is defined on $[0,1]$ by

$$Q(t) = F_X^{-1}(t) = \inf\{x \in \mathcal{I}_X : F_X(x) \geq t\},$$

it is right-continuous with left-hand limits, like the distribution function. For every uniform variable U, $F_X^{-1}(U)$ has the distribution function F_X and, if F is continuous, then $F(X)$ has an uniform distribution function. The inverse of the distribution function satisfies $F_X^{-1} \circ F_X(x) = x$ for every x in the support of X and $F_X \circ F_X^{-1} = id$ for every continuity point x of F_X. The weak convergence of the empirical uniform process and its functionals have been widely studied (Shorack and Wellner, 1986, van der Vaart and Wellner, 1996). For a differentiable functional $\psi(F_X)$, $n^{1/2}\{\psi(\widehat{F}_{X,n}) - \psi(F_X)\}$ converges weakly to $(\psi' B) \circ F_X$ where B is a Brownian motion, limiting distribution of the empirical process $n^{1/2}(\widehat{F}_{X,n} - F_X)$. It follows that the process $n^{1/2}(\widehat{F}_{X,n}^{-1} - F_X^{-1})$ converges weakly to $B \circ F_X (f_X \circ F_X)^{-1}$. Kiefer (1972) established a law of iterated logarithms for quantiles of probabilities tending to zero, the same result holds for $1 - p_n$ as p_n tend to one.

Theorem 1.2 (Kiefer 1972). *Let p_n tend to zero with np_n and h_n tend to infinity, and let $\delta = 1$ or -1 then*

$$\limsup_n \delta \frac{Q_n(p_n) - np_n}{\{2np_n \log\log n\}^{1/2}} = 1, a.s.$$

The results were extended to conditional distribution functions and Sheather and Marron (1990) considered kernel quantile estimators. The inverse function for a nonparametric regression curve determines thresholds for X given Y values, it is related to the distribution function of Y conditionally on X. The inverse empirical process for a monotone nonparametric regression function has been studied in Pinçon and Pons (2006) and Pons (2008), the main results are presented and generalized in Chapter 5. The behaviour of the threshold estimators $\widehat{Q}_{X,n,h}$ and $\widehat{Q}_{Y,n,h}$ of the conditional distribution is studied, with their bias and variance and the mean squared errors which determine the optimal bandwidths specific to the quantile processes.

The Bahadur representation for the quantile estimators is an expansion

$$\widehat{F}^{-1}_{X,n}(t) = F^{-1}_X(t) + \frac{t - \widehat{F}_{X,n}}{f_X} \circ F^{-1}_X(t) + R_n(t), \ t \in [0,1],$$

where the main is a sum of independent and identically distributed random variables and the remainder term $R_n(t)$ is a $o_p(n^{-1/2})$ (Ghosh, 1971), Bahadur (1966) studied its *a.s.* convergence. Lo and Singh (1986), Gijbels and Veraverbeke (1988, 1989) extended this approach by differentiation to the Kaplan-Meier estimator of the distribution function of independent and identically distributed right-censored variables.

1.3 Estimation of functionals of processes

Watson and Laedbetter (1964) introduced smooth estimators for the hazard function of a point process. The functional intensity $\lambda(t)$ of an inhomogeneous Poisson point process N is defined by

$$\lambda(t) = \lim_{\delta \to 0} \delta^{-1} P\{N(t+\delta) - N(t^-) = 1 \mid N(t-)\},$$

it is estimated using a kernel smoothing, from the sample-path of the point process observed on an interval $[0,T]$. Let $Y(t) = N(T) - N(t)$, then

$$\widehat{\lambda}_h(t) = \int K_h(t-s) 1_{\{Y(s)>0\}} Y^{-1}(s)\, dN(s).$$

For a sample of a time variable T with distribution function F, let $\bar{F}$ be the survival function of the variable T, $\bar{F} = 1 - F^-$, the hazard function λ is now defined as $\lambda(t) = f(t)/\bar{F}(t)$. The probability of excess is

$$P_t(t+x) = \Pr(T > t + x \mid T > t) = 1 - \frac{F(t+x) - F(t)}{\bar{F}(t)}$$
$$= \exp\{-\int_t^{t+x} \lambda(s)\, ds\}.$$

The product-limit estimator has been defined for the estimation of the distribution function of a time variable under an independent right-censorship by Kaplan and Meier (1957). Breslow and Crowley (1974) studied the asymptotic behaviour of the process $B_n = n^{1/2}(\widehat{F}_n - F)$, they proved its weak convergence to a Gaussian process B with independent increments, mean zero and a finite variance on every compact sub-interval of $[0, T_{n:n}]$, where $T_{n:n} = \max_{i\leq n} T_i$. The weak convergence of $n^{1/2}(\bar{F}_n - \bar{F})$ has been extended by Gill (1983) to the interval $[0, T_{n:n}]$ using its expressions as a martingale up to the stopping time $T_{n:n}$. Let $\tau_F = \sup\{t; F(t) < 1\}$, for $t < \tau_F$ and if $\int_0^{\tau_F} \bar{F}^{-1}\, d\Lambda < \infty$, we have

$$\widehat{\Lambda}_n(t) = \int_0^{t\wedge T_{n:n}} \frac{d\widehat{F}_n(s)}{1 - \widehat{F}_n^-(s)},$$
$$\widehat{F}_n(t) = \int_0^{t\wedge T_{n:n}} \widehat{\bar{F}}_n(s)\, d\widehat{\Lambda}_n(s),$$
$$\frac{F - \widehat{F}_n}{1 - F}(t) = \int_0^{t\wedge T_{n:n}} \frac{1 - \widehat{F}_n(s^-)}{1 - F(s)} \{d\widehat{\Lambda}_n(s) - d\Lambda(s)\}$$

as a consequence, the process $n^{1/2}(F - \widehat{F}_n)\bar{F}^{-1}$ converges weakly on $[0, \tau_F[$ to a centered Gaussian process B_F, with independent increments and variance $v_{\bar{F}}(t) = \int_0^t \{(1-F)^{-1}\bar{F}\}^2\, dv_\Lambda$, where v_Λ is the asymptotic variance of the process $n^{1/2}(\widehat{\Lambda}_n - \Lambda)$.

The definition of the intensity is generalized to point processes having a random intensity. For a multiplicative intensity λY, with a predictable process Y, the hazard function λ is estimated by

$$\widehat{\lambda}_h(t) = \int K_h(t-s) Y^{-1}(s) 1_{\{Y(s)>0\}}\, dN(s).$$

For a random time sample $(T_i)_{i\leq n}$, $N(t) = \sum_{i=1}^n 1_{\{T_i \leq t\}}$ and the process Y is $Y(t) = \sum_{i=1}^n 1_{\{T_i \geq t\}}$. Under a right-censorship of a time variable T by an independent variable C, only $T \wedge C$ and the indicator δ of the event $\{T \leq C\}$ are observed. Let $X = T \wedge C$, the counting processes for a n-sample of (X, δ) are $N(t) = \sum_i 1_{\{T_i \leq t \wedge C_i\}}$ and $Y(t) = \sum_i 1_{\{X_i \geq t\}}$. Martingale techniques are used to expand the estimation errors, providing optimal convergence rates according to the regularity conditions for the hazard function (Pons, 1986) and the weak convergences with fixed or variable bandwidths (Chapter 6). Regression models for the intensity are classical, there have generally the form $\lambda(t; \beta) = \lambda(t) r_\beta(Z(t))$ with a regressor process $(Z(t))_{t \geq 0}$ and a parametric regression function such as $r_\beta(Z(t)) = r(\beta^T Z(t))$, with an exponential function r in the Cox model (1972). The classical estimators of the Cox model rely on the estimation of the cumulated hazard function $\Lambda(t) = \int_0^t \lambda(s)\, ds$ by the stepwise process $\widehat{\Lambda}_n(t; \beta)$ at fixed β and the parameter β of the exponential regression function $r_Z(t; \beta) = e^{\beta^T Z(t)}$ is estimated by maximization of an expression similar to the likelihood where λ is replaced by the jump of $\widehat{\Lambda}_n(\beta)$ at T_i (Cox 1972).

The asymptotic properties of the estimators for the cumulated hazard function and the parameters of the Cox model were established by Andersen and Gill (1982), among others. The estimators presented in this chapter are obtained by minimization of partial likelihoods based on kernel estimators of the baseline hazard function λ defined for each model and on histogram estimators. In the multiplicative intensity model, the kernel estimator of λ satifies the same minimax property as the kernel estimator of a density (Pons, 1986) and this property is still satisfied in the multiplicative regression models of the intensity. Pons and Turckheim (1987) proved the asymptotic equivalence of the estimators of an exponential regression model based on the estimated cumulative intensity and a histogram estimator. The comparison is extended to the new estimators defined from kernel estimators of hazard functions in this book.

For a spatial stationary process N on $\mathbb{R}^d$, the k-th moment measures defined for $k \geq 2$ and for every continuous and bounded function g on $(\mathbb{R}^d)^k$ by

$$\nu_k(g) = E \int_{(\mathbb{R}^d)^k} g(x_1, \ldots, x_k) N(dx_1) \ldots N(dx_k)$$

have been intensively studied and they are estimated by empirical moments from observations on a subset G of $\mathbb{R}^d$. The centered moments are immediatly obtained from the mean measure m and $\mu_k = \sum_{i=1}^k (-1)^i C_k^i m^i \nu_{k-i}$. The stationarity of the process implies that the k-th moment of N

is expressed as the expectation of an integral of a translation of its $(k-1)$-th moment

$$\nu_k(g) = E \int_{(\mathbb{R}^d)^k} g(x_1 - x_k, \ldots, x_{k-1} - x_k, 0) N(dx_1) \ldots N(dx_k)$$

which develops in the form

$$\begin{aligned}\nu_k(g) &= E \int_{\mathbb{R}^d} \{ \int_{(\mathbb{R}^d)^{k-1}} g_{k-1} \circ T_x(x_1, \ldots, x_{k-1}) N(dx_1) \ldots N(dx_{k-1}) \} \, N(dx) \\ &= E \int_{\mathbb{R}^d} \nu_{k-1}(g_{k-1} \circ T_x) \, N(dx),\end{aligned}$$

where $g_{k-1}(x_1, \ldots, x_{k-1}) = g(x_1, \ldots, x_{k-1}, 0)$. Let λ be the Lebesgue measure on $\mathbb{R}^d$ and T_x be the translation operator of x in $\mathbb{R}^d$, then the moment estimators are built iteratively by the relationship

$$\widehat{\nu}_{k,G}(g) = \{\lambda(G)\}^{-1} \int_G \widehat{\nu}_{k-1,G^{k-1}}(g_{k-1} \circ T_y) dN(y).$$

The estimator is consistent and its convergence rate is $\{\lambda(G)\}^{k/2}$. The stationarity of the process and a mixing condition imply that for every function g of $C_b((\mathbb{R}^d)^k)$, the variable $\{\lambda(G)\}^{k/2}(\widehat{\nu}_{k,G}(g) - \nu_k(g))$ converges weakly to a normal variable with variance $\nu_{2k}(g)$. The density of the k-th moment measures are defined as the derivatives of ν_k with respect to the Lebesgue measure on $\mathbb{R}^d$ and they are estimated by smoothing the empirical estimator $\widehat{\nu}_{k,G}$ using a kernel K_h on $\mathbb{R}^d$ and, by iterations, on $\mathbb{R}^{kd}$. The convergence of the kernel estimator is then $h^{kd/2}\{\lambda(G)\}^{k/2}$, as a consequence of the k d-dimensional smoothing.

Consider a diffusion model with nonparametric drift function α and variance function, or diffusion, β

$$dX_t = \alpha(X_t)dt + \beta(X_t)dB_t, t \geq 0 \tag{1.11}$$

where B is the standard Brownian motion. The drift and variance are expressed as limits of variations of X

$$\begin{aligned}\alpha(X_t) &= \lim_{h \to 0} h^{-1} E\{(X_{t+h} - X_t) \mid X_t\}, \\ \beta(X_t) &= \lim_{h \to 0} h^{-1} E\{(X_{t+h} - X_t)^2 \mid X_t\}.\end{aligned}$$

The process X can be approximated by nonparametric regression models with regular or variable discrete sampling schemes of the sample-path of the process X. The diffusion equation uniquely defines a continuous process $(X_t)_{t>0}$. Assuming that $E\exp\{-\frac{1}{2}\int_0^t \beta^2(B_s)\,ds\}$ is finite, the Girsanov theorem formulates the density of the process X. Parametric diffusion models have been much studied and estimators of the parameters are defined by maximum likelihood from observations at regularly spaced discretization points or at random stopping times. In a discretization scheme with a constant interval of length Δ_n between observations, nonparametric estimators are defined like with samples of variables in nonparametric regression models (Pons, 2008). Let $(X_{t_i}, Y_i)_{i\leq 1}$ be discrete observations with $Y_i = X_{t_{i+1}} - X_{t_i}$ defined by equation (1.11), the functions α and β^2 are estimated by

$$\widehat{\alpha}_n(x) = \frac{\sum_{i=1}^n Y_i K_{h_n}(x - X_{t_i})}{\Delta_n \sum_{i=1}^n K_{h_n}(x - X_{t_i})},$$

$$\widehat{\beta}_n^2(x) = \frac{\sum_{i=1}^n Z_i^2 K_{h_n}(x - X_{t_i})}{\Delta_n \sum_{i=1}^n K_{h_n}(x - X_{t_i})},$$

where $Z_i = Y_i - \Delta_n \widehat{\alpha}_n(X_{t_i})$ is the variable of the centered variations for the diffusion process. The variance of the variable Y_i conditionally on X_{t_i} varies with X_{t_i} and weighted estimators are also defined here. Varying sampling intervals or random sampling schemes modify the estimators. Functional models of diffusions with discontinuities were also considered in Pons (2008) where the jump size was assumed to be a squared integrable function of the process X and a nonparametric estimator of this function was defined. Here the estimators of the discretized process are compared to those built with the continuously observed diffusion process X defined by (1.11), on an increasing time interval $[0, T]$. The kernel bandwidth h_T tends to zero as T tends to infinity with the same rate as h_n. In Chapter 8, the MISE of each estimator and its optimal bandwidth are determined. The estimators are compared with those defined for the continuously observed diffusion processes.

Nonparametric time and space transformations of a Gaussian process have been first studied by Perrin (1999), Guyon and Perrin (2000) who estimated the function Φ of non-stationary processes $Z = X \circ \Phi$, with X a stationary Gaussian process, Φ a monotone continuously differentiable function defined in $[0,1]$ or in $[0,1]^3$. The covariance of the process Z is $r(x,y) = R(\Phi(x) - \Phi(y))$ where R is the stationary covariance of X, which implies $R(u) = r(0, \Phi^{-1}(u))$ and $R(-u) = R(u)$, with a singularity at zero. The singularity function of Z is the difference $\xi(x)$ of the left and right derivatives of $r(x,x)$, which implies $\Phi(x) = v^{-1}(1)v(x)$ where $v(x)$ equals

$\int_0^x \xi(u)\,du$. The estimators are based on the covariances of the process Z are built with its quadratic variations. For the time transformation, the estimator of $\Phi(x)$ is defined by linearisation of $V_n(x) = \sum_{k=1}^{[nx]}(\Delta Z_k)^2$ where the variables $Z_k = Z(n^{-1}k) - Z(n^{-1}(k-1))$ are centered and independent

$$\begin{aligned} v_n(x) &= V_n(x) + (nx - [nx])(\Delta Z_{[nx]+1})^2,\ x \in [0,1[,\\ v_n(1) &= V_n(1), \\ \widehat{\Phi}_n(x) &= v_n^{-1}(1)v_n(x), \end{aligned} \tag{1.12}$$

the process $\widehat{\Phi}_n - \Phi$ is uniformly consistent and $n^{1/2}(\widehat{\Phi}_n - \Phi)$ is asymptotically Gaussian. The method was extended to $[0,1]^3$. The diffusion processes cannot be reduced to the same model but the method for estimating its variance function relies on similar properties of Gaussian processes.

In time series analysis, the models are usually defined by scalar parameters and a wide range of parametric models for stationary series have been intensively studied since many years. Nonparametric spectral densities of the parametric models have been estimated by smoothing the periodogram calculated from T discrete observations of stationary and mixing series (Wold, 1975, Brillinger, 1981, Robinson, 1986, Herrmann, Gasser and Kneip, 1992, Pons, 2008). The spectral density is supposed to be twice continuously differentiable and the bias, variance and moments of its kernel estimator have been expanded like those of a probability density. It converges weakly with the rate $T^{2/5}$ to a Gaussian process, as a consequence of the weak convergence of the empirical periodogram.

1.4 Content of the book

In each chapter, the classical estimators for samples of independent and identically distributed variables are presented, with approximations of their bias, variance and L_p-moments, as the sample size n tends to infinity and the bandwidth to zero. In each model, the weak convergence of the whole processes are considered and the limiting distributions are not centered for the optimal bandwidth minimizing the mean integrated squared error.

Chapters 2 and 3 focus on the density and the regression models, respectively. In models with a constant variance, the regression estimator defined as a ratio of kernel estimators is approximated by a weighted sum of two kernel estimators and its properties are easily deduced. In models with a functional variance, a kernel estimator of the variance is also considered and the estimator of the regression function is modified by an empirical weight. The properties of the modified estimator are detailed. The estimators for independent and identically distributed variables are extended to a stationary continuous process $(X_t)_{t\geq 0}$ continuously observed on an increasing time interval, for the estimation of the ergodic density of the process. The observations at times s and t are dependent so the methods for independent observations do not adapt immediatly. The estimators are defined with the conditions necessary for their convergences and their approximation properties are proved. The optimal bandwidth minimizing the mean squared error are functional sequences of bandwidths and the properties of the kernel estimators are extended to varying bandwidths for this reason in Chapter 4.

The estimators of derivatives of the density, regression function and the other functions are expressed by the means of derivatives of the kernel so that their convergence rate is modified, the k-th derivative of K_h being normalized by $h^{-(k+1)}$ instead of h^{-1} for K_h. Functionals of the densities and functions in the other models are considered, the asymptotic properties of their estimators are deduced from those of the kernel estimators.

The inverse function defined for the increasing distribution function F are generalized in Chapter 5 to conditional distribution functions and to monotone regression functions. The bias, variance, norms, optimal bandwidths and weak convergences of the quantiles of their kernel estimators are established with detailed proofs. Exact Bahadur-type representations are written, with L_2 approximations.

Chapter 6 provides new kernel estimators in nonparametric models for real point processes which generalize the martingale estimators of the baseline hazard functions already studied. They are compared to new histogram-type estimators built for these functional models. The probability density of excess duration for a point process and its estimator are defined and the properties of the estimator are also studied.

The single-index models are nonparametric regression models for linear combinations of the regression variables. The estimators of the parameter vector θ and the nonparametric regression function g of the model are proved to be consistent for independent and identically distributed variables. New estimators of g and θ are considered in Chapter 7, with direct estimation methods, without numerical iteration procedures. The convergence rate for the estimator $\widehat{\theta}_{n,h}$ obtained by minimizing the empirical mean squared estimation error $\widehat{V}_n$ is $(nh^3)^{1/2}$. The estimator $\widehat{m}_{n,h}$ built with this estimator of θ has the same convergence rate which is not so small as the nonparametric regression estimator with a d-dimensional regression variable. A differential empirical squared error criterion provides an estimator for the parameter which converges more quickly and the estimator of the regression function m has the usual nonparametric convergence rate $(nh)^{1/2}$. More generally, the linear combination of the regressors can be replaced by a parametric change of variable, in a regression model $Y = g \circ \varphi_\theta(X) + \varepsilon$. Replacing the function g by a kernel estimator at fixed θ, the parameter in then estimated by minimizing an empirical version of the error $V(\theta) = \{Y - \widehat{g}_{n,h} \circ \varphi_\theta(X)\}^2$. Its asymptotic properties are similar to those of the single-index model estimators. The optimal bandwidths are precised.

The estimators of the drift and variance of continuous diffusion processes depend on the sampling scheme for their discretization and they are compared to the estimators built from the whole sample-path of the diffusion process. New results are presented in Chapter 8 and they are extended to the sum of a diffusion processes and a jump process governed by the diffusion. For nonstationary Gaussian models, a kernel estimator is defined for the singularity function of the covariance of the process.

In Chapter 9, classical estimators of covariances and nonparametric regression functions used for stationary time series are generalized to nonstationary models. The expansions of the bias, variance and L_p-errors are detailed and optimal bandwidths are defined. Nonparametric estimators are defined for the stationarization of time series and for their mean function in auto-regressive models, based on the results of the previous chapters.

Chapter 2

Kernel estimator of a density

2.1 Introduction

Let f be the continuous probability density of a real variable X defined on a probability space $(\Omega, \mathcal{A}, P)$ and F be its distribution functions. Let $\mathcal{I}_X$ be the finite or infinite support of the density function f of X with respect to the Lebesgue measure and $\mathcal{I}_{X,h} = \{s \in \mathcal{I}_X; [s-h, s+h] \in \mathcal{I}_X\}$. For a sample $(X_i)_{1 \le i \le n}$ distributed as X and a kernel K, estimators of F and f are defined on $\Omega \times \mathbb{R}$ as the empirical distribution function

$$\widehat{F}_{X,n}(x) = n^{-1} \sum_{i=1}^{n} 1_{\{X_i \le x\}},\ x \in \mathcal{I}_X$$

and the kernel estimator is defined for every x in $\mathcal{I}_{X,h}$ as

$$\widehat{f}_{X,n,h}(x) = \int K_h(x-s)\, d\widehat{F}_{X,n}(s) = \frac{1}{n} \sum_{i=1}^{n} K_h(x - X_i),$$

where $K_h(x) = h^{-1}K(h^{-1}x)$ and $h = h_n$ tends to zero as n tends to infinity and 1_A is the indicator of a set A. The empirical probability measure is $\widehat{P}_{X,n,h}(A) = n^{-1} \sum_{i=1}^{n} \delta_{X_i}(A)$, with $\delta_{X_i}(A) = 1_{\{X_i \in A\}}$. Let

$$f_{n,h}(x) = E\widehat{f}_{n,h}(x) = \int K_h(x-s)\, dF(s),$$

the bias of the kernel estimator $\widehat{f}_{n,h}(x)$ is

$$b_{n,h}(x) = f_{n,h}(x) - f(x) = \int K(t)\{f(x+ht) - f(x)\}\, dt. \tag{2.1}$$

The L_p-risk of the kernel estimator of the density f of X is its L_p-norm

$$\|\widehat{f}_{n,h}(x) - f(x)\|_p = \{E|\widehat{f}_{n,h}(x) - f(x)|^p\}^{1/p} \tag{2.2}$$

and it is bounded by the sum of a p-moment and a bias term. For every x in $\mathcal{I}_{X,h}$, the pointwise and uniform convergence of the kernel estimator $\widehat{f}_{n,h}$ are established under the following conditions about the kernel and the density.

Condition 2.1.

(1) K is a symmetric density such that $|x|^2 K(x) \to 0$ as $|x|$ tends to infinity or K has a compact support with value zero on its frontier;
(2) The density function f belongs to the class $C_2(I_X)$ of twice continuously differentiable functions defined in I_X.
(3) The kernel function satisfies integrability conditions: the moments $m_{2K} = \int u^2 K(u)du$, $\kappa_\alpha = \int K^\alpha(u)du$, for $\alpha \geq 0$, and $\int |K'(u)|^\alpha du$, for $\alpha = 1, 2$, are finite. As $n \to \infty$, $h_n \to 0$ $nh_n \to \infty$.
(4) nh_n^5 converges to a finite limit γ.

The next conditions are stronger than Conditions 2.1 (2)-(4), with higher degrees of differentiability and integrability.

Condition 2.2.

(1) The density function f is $C_s(I_X)$, with a continuous and bounded derivative of order s, $f^{(s)}$, on $\mathcal{I}_X$.
(2) As $n \to \infty$, $nh_n \to 0$ and nh_n^{2s+1} converges to a finite limit $\gamma > 0$. The kernel function satifies $m_{jK} = \int u^j K(u)du = 0$ for $j < s$, m_{sK} and $\int |K'(u)|^\alpha du$ are finite for $\alpha \leq s$.

The conditions may be strengthened to allow a faster rate of convergence of the bandwidth to zero by replacing the strictly positive limit of nh_n^{2s+1} by $nh_n^{2s+1} = o(1)$. That question appears crucial in the relative importance between the bias and the variance in the L_2-risk of $\widehat{f}_{n,h} - f$. The choice of the optimal bandwidth minimizing that risk corresponds to an equal rate for the squared bias and the variance and implies the rates of Condition 2.1(4) or 2.2(2) according to the derivability of the density. Considering the normalized estimator, the reduction of the bias requires a faster convergence rate.

2.2 Risks and optimal bandwidths for the kernel estimator

Proposition 2.1. *Under Conditions 2.1-1 for a continuous density f, the estimator $\widehat{f}_{n,h}(x)$ converges in probability to $f(x)$, for every x in $I_{X,h}$. Moreover, $\sup_x |\widehat{f}_{n,h}(x) - f(x)|$ tends a.s. to infinity as n tends to infinity if and only if f is uniformly continuous.*

Proof. The first assertion is a consequence of an integration by parts

$$\sup_{x\in\mathcal{I}_{X,h}} |\widehat{f}_{n,h}(x) - f_{n,h}(x)| \le \sup_{x\in\mathcal{I}_{X,h}} \frac{1}{h}\int |\widehat{F}_{n,h}(y) - F(y)|\,|dK(\frac{x-y}{h})|$$
$$\le \frac{1}{h}\sup_y |\widehat{F}_{n,h}(y) - F(y)| \int |dK|.$$

The Dvoretzky, Kiefer and Wolfowitz (1956) exponential bound implies that for every $\lambda > 0$, $\Pr(\sup_{\mathcal{I}_X} n^{1/2}|\widehat{F}_{n,h} - F| > \lambda) \le 58\exp\{-2\lambda^2\}$, then

$$\Pr(\sup_{\mathcal{I}_{X,h}} |\widehat{f}_{n,h} - f_{n,h}| > \varepsilon) \le \Pr(\sup_{\mathcal{I}_X} |\widehat{F}_{n,h} - F| > (\int |dK|)^{-1} h_n\varepsilon)$$
$$\le 58\exp\{-\alpha n h_n^2\}$$

with $\alpha > 0$, and $\sum_{n=1}^{\infty} \exp\{-n\alpha\, h_n^2\}$ tends to zero under Condition 2.1 or 2.2. □

Proposition 2.2. *Assume $h_n \to 0$ and $nh_n \to \infty$,*
(a) under Conditions 2.1, the bias of $\widehat{f}_{n,h}(x)$ is

$$b_{n,h}(x) = \frac{h^2}{2} m_{2K} f^{(2)}(x) + o(h^2),$$

denoted $h^2 b_f(x) + o(h^2)$, its variance is

$$Var\{\widehat{f}_{n,h}(x)\} = (nh)^{-1}\kappa_2\, f(x) + o((nh)^{-1}),$$

also denoted $(nh)^{-1}\sigma_f^2(x) + o((nh)^{-1})$, where all approximations are uniform. Let K have the compact support $[-1,1]$, the covariance of $\widehat{f}_{n,h}(x)$ and $\widehat{f}_{n,h}(y)$ is zero if $|x-y| > 2h$, otherwise it is approximated by

$$\frac{(nh)^{-1}}{2}\{f(x) + f(y)\}\delta_{x,y}\int K((v - \alpha_h)K(v + \alpha_h)dv$$

where $\alpha_h = |x-y|/(2h)$ and $\delta_{x,y}$ is the indicator of $\{x = y\}$.

(b) Under Conditions 2.2, for every $s \ge 2$, the bias of $\widehat{f}_{n,h}(x)$ is

$$b_{n,h}(x;s) = \frac{h^s}{s!} m_{sK} f^{(s)}(x) + o(h^s),$$

and

$$\|\widehat{f}_{n,h}(x) - f_{n,h}(x)\|_p = 0((nh)^{-1/p}),$$

for every $p \ge 2$, where the approximations are uniform.

Proof. The bias as h tends to zero is obtained from a second order expansion of $f(x+ht)$ under Condition 2.1, and from its s-order expansion under Condition 2.2. The variance of $\widehat{f}_{n,h}(x)$ is

$$Var\{\widehat{f}_{n,h}(x)\} = n^{-1}\{\int K_h^2(x-s)f(s)\,ds - f_{n,h}^2(x)\}.$$

The first term of the sum is $n^{-1}\int K_h^2(x-u)f(u)du = (nh)^{-1}\kappa_2 f(x) + o((nh)^{-1})$, the second term $n^{-1}f^2(x) + O(n^{-1}h)$ is smaller.

The covariance of $\widehat{f}_{n,h}(x)$ and $\widehat{f}_{n,h}(y)$ is written $n^{-1}\{\int_{\mathcal{I}_X^2} K_h(u-x)K_h(u-y)f(u)\,du - f_{n,h}(x)f_{n,h}(y)\}$, it is zero if $|x-y|>2h$. Otherwise let $\alpha_h = |x-y|/(2h) < 1$, changing the variables as $h^{-1}(x-u) = v-\alpha_h$ and $h^{-1}(y-u) = v+\alpha_h$ with $v = \{(x+y)/2-u\}/h$, the covariance develops as

$$Cov\{\widehat{f}_{n,h}(x),\widehat{f}_{n,h}(y)\} = (nh)^{-1}f(\frac{x+y}{2})\int K(v-\alpha_h)K(v+\alpha_h)dv + o((nh)^{-1}).$$

If $|x-y|\le 2h$, $f((x+y)/2) = f(x)+o(1) = f(y)+o(1)$, the covariance is approximated by

$$\frac{(nh)^{-1}}{2}\{f(x)+f(y)\}\mathcal{I}_{\{0\le\alpha_h<1\}}\int K\big((v-\alpha_h\big)K\big(v+\alpha_h\big)dv.$$

Due to the compactness of the support of K, the covariance is zero if $\alpha_h \ge 1$. For $x\ne y$, α_h tends to infinity and $I\{0\le\alpha_h<1\}$ tends to zero as n tends to infinity, then the indicator is approximated by the indicator $\delta_{x,y}$ of $\{x=y\}$.

For $p=1$, $E|\widehat{f}_{n,h} - f_{n,h}|(x) \le \int_{\mathcal{I}_{X,h}} |K_h(x-s)|\,d|\widehat{F}_n - F|(s)$ which converges to zero as n tends to infinity. For $p\ge 3$, the L_p-risk of $\widehat{f}_{n,h}(x)$ is obtained from the expansion of the sum and by recursion on the order of the moment in the expansion of

$$E\{\widehat{f}_{n,h}(x) - f_{n,h}(x)\}^p = E\big[n^{-1}\sum_{i=1}^n\{K_h(x-X_i) - f_{n,h}(x)\}\big]^p.$$

The first moments of the centered estimator $\widehat{f}_{n,h} - f_{n,h}$ are

$$E\{\widehat{f}_{n,h}(x) - f_{n,h}(x)\}^3 = (nh)^{-1}\{-2\kappa_2 f^2(x) + o(1)\},$$
$$E\{\widehat{f}_{n,h}(x) - f_{n,h}(x)\}^4 = (nh)^{-1}\{\kappa_2 f^3(x) + o(1)\},$$
$$E\{\widehat{f}_{n,h}(x) - f_{n,h}(x)\}^5 = (nh)^{-1}\{11\kappa_2 f^3(x) + o(1)\}.$$

By iterations, the product of moments in the expansion of the risk $E\{\widehat{f}_{n,h}(x) - f_{n,h}(x)\}^p$ determines its higher order as $(nh)^{-1}$. □

Integrating the above expansions entails similar bounds for the integrated norms $E\int|\widehat{f}_{n,h}(x)-f_{n,h}(x)|^p\,dx = 0((nh)^{-1})$, for every $p>1$.

For $p=2$, $E\{\widehat{f}_{n,h}(x)-f(x)\}^2 = Var\{\widehat{f}_{n,h}(x)\}+\{f_{n,h}(x)-f(x)\}^2$ and its first order expansion is $n^{-1}h^{-1}\kappa_2 f(x)+o(n^{-1}h^{-1})+\frac{1}{4}m_{2K}^2h^4f^{(2)2}(x)+o(h^4)$. The asymptotic mean squared error for $\widehat{f}_{n,h}$ at x is then

$$AMSE(\widehat{f}_{n,h};x) = (nh)^{-1}\kappa_2 f(x)+\frac{1}{4}m_{2K}^2h^4f^{(2)2}(x),$$

it is minimum for the bandwidth function

$$h_{AMSE}(x) = n^{-1/5}\{\frac{\kappa_2 f(x)}{m_{2K}^2 f^{(2)2}(x)}\}^{1/5}.$$

A smaller order bandwidth increases the variance of the density estimator and reduces its bias, with the order $n^{-1/5}$ its asymptotic distribution cannot be centered. An estimator of the derivative $f^{(k)}$ is defined by the means of the derivative $K^{(k)}$ of the symmetric kernel, for $k\geq 1$. The convergences rates for estimators of a derivative of the density also depend on the order of the derivative. Consider the k-order derivative of K_h

$$K_h^{(k)}(x) = h^{-(k+1)}K^{(k)}(h^{-1}x),\ k\geq 1.$$

The estimators of the derivatives of the density are

$$\widehat{f}_{n,h}^{(k)}(x) = n^{-1}\sum_{i=1}^n K_h^{(k)}(x-X_i). \tag{2.3}$$

The next lemma implies the uniform consistency of $\widehat{f}_{n,h}^{(k)}$ to $f^{(k)}$, for every order of derivability $k\geq 1$ and allows to calculate the variance of the derivative estimators. It is not exhaustive and integrals of higher orders are easily obtained using integrations by parts.

Lemma 2.1. *Let K be a symmetric density function in class C_2, its derivatives satisfy the following properties : $\int K^{(j)}(z)\,dz = 0$, for every $j\geq 1$, $\int zK^2(z)\,dz = 0$, $\kappa_{22} = \int z^2K^2(z)\,dz \neq 0$ and*

$$\int zK^{(1)}(z)\,dz = -1,\ \int z^2K^{(1)}(z)\,dz = 0,\ \int z^3K^{(1)}(z)\,dz = -3m_{2K},$$

$$\int zK^{(2)}(z)\,dz = 0,\ \int z^2K^{(2)}(z)\,dz = 2,\ \int z^3K^{(2)}(z)\,dz = 0,$$

$$\int z^4K^{(2)}(z)\,dz = 12m_{2K},\ \kappa_{11} = \int z(K'K)(z)\,dz = -\kappa_2/2,$$

$$\int K^{(1)}K\,dz = 0,\ \int K^{(1)2}\,dz \neq 0.$$

The sum $\widehat{f}^{(1)}_{n,h}(x) = n^{-1}\sum_{i=1}^n K_h^{(1)}(x - X_i)$ converges uniformly on $\mathcal{I}_{X,h}$ to its expectation

$$\begin{aligned}
f^{(1)}_{n,h}(x) &= EK_h^{(1)}(x - X) = \int K_h^{(1)}(u - x) f_X(u)\,du \\
&= -f^{(1)}(x)\int zK^{(1)}(z)\,dz - \frac{h^2}{6} f^{(3)}(x)\int z^3K^{(1)}(z)\,dz + o(h^2) \\
&= f^{(1)}(x) + \frac{h^2}{2} m_{2K} f^{(3)}(x) + o(h^2),
\end{aligned}$$

then $\widehat{f}^{(1)}_{n,h}$ converges uniformly to $f^{(1)}(x)$ and its bias is $\frac{h^2}{2} m_{2K} f^{(3)}(x)$. Its variance is $(nh^3)^{-1} f(x)\int K^{(1)2}(z)\,dz + o((nh^3)^{-1})$ and the optimal local bandwidth for estimating $f^{(1)}$ is deduced as

$$h_{AMSE}(f^{(1)}; x) = n^{-1/7}\Big\{\frac{f(x)\int K^{(1)2}(z)\,dz}{m_{2K}^2 f^{(3)2}(x)}\Big\}^{1/7},$$

thus the estimator of the first density derivative (2.3) has to be computed with a bandwidth estimating $h_{AMSE}(f^{(1)}; x)$. For the second derivative, the expectation of $\widehat{f}^{(2)}_{n,h}$ is $f^{(2)}_{n,h}(x) = f^{(2)}(x) + \frac{h^2}{2} m_{2K} f^{(3)}(x) + o(h^2)$, so it converges uniformly to $f^{(2)}$ with the bias $\frac{h^2}{2} m_{2K} f^{(4)}(x) + o(h^2)$ and the variance $(nh^5)^{-1} f(x)\int K^{(2)2}(z)\,dz + o((nh^4)^{-1})$. More generally, Lemma 2.1 generalizes by induction to higher orders and the rate of optimal bandwidths is deduced as follows.

Proposition 2.3. *Under Conditions 2.1, the estimator $\widehat{f}^{(k)}_{n,h}$ of the k-order derivative of a density in class C_2 has a bias $O(h^2)$ and a variance $O((nh^{2k+1})^{-1})$, its optimal local and global bandwidths are $O(n^{-1/(2k+5)})$, for every $k \geq 2$.*
For a density of class C_s and under Conditions 2.2, the bias is a $O(h^s)$ and the variance a $O((nh^{2k+1})^{-1})$, its optimal bandwidths are $O(n^{-1/(2k+2s+1)})$ and the corresponding L_2-risks are $O(n^{-s/(2k+2s+1)})$.

As a consequence the L_2-risk of the estimator $\widehat{f}^{(k)}_{n,\widehat{h}_{opt}}$ is a $O(n^{-2s/(2k+2s+1)})$ for every density in C_s, $s \geq 2$. If the k-th derivative of the kernel and the density are lipschitzian with $|K^{(k)}(x) - K^{(k)}(y)| \leq \alpha|x - y|$ and $|f^{(k)}(x) - f^{(k)}(y)| \leq \alpha|x - y|$ for some constant $\alpha > 0$, then there exists a constant C such that for every x and y in $\mathcal{I}_{X,h}$

$$|\widehat{f}^{(k)}_{n,h}(x) - \widehat{f}^{(k)}_{n,h}(y)| \leq C\alpha h^{-(k+1)}|x - y|.$$

The integral $\theta_k = \int f^{(k)2}(x)\,dx$ of the quadratic k-th derivative of the density is estimated by

$$\widehat{\theta}_{k,n,h} = \int \widehat{f}^{(k)2}_{n,h}(x)\,dx, \tag{2.4}$$

the variance $E(\widehat{\theta}_{k,n,h} - \theta_k)^2$ has the same order as the MISE for the estimator $\widehat{f}^{(k)}_{n,h}$ of $f^{(k)}$, hence it converges to θ_k with the rate $O((n^{1/2}h^{k+1/2})$ and the estimator does not achieve the parametric rate of convergence $n^{1/2}$.

The L_p-risk of the estimator of the density decreases as s increases and, for $p \geq 2$, a bound of the L_p-norm is

$$\|\widehat{f}_{n,h}(x) - f(x)\|_p^p \leq 2^{p-1}\Big(\frac{h^{ps}}{(s!)^p}\{m_{sK}^p f^{(k)p}(x) + o(1)\} \\ + (nh)^{-1}\{g_p(x) + o(1)\}\Big),$$

where $g_p(x) = \sum_{k=2}^{[p/2]} \sum_{1<j_1 \neq \ldots \neq j_k \leq p; \sum_i j_i = p} \kappa_{j_1} \ldots \kappa_{j_l} f^k(x)$. The optimal bandwidth is still reached when both terms of this bound are of the same order and minimal. With $p = 2$ and $s = 2$, it is

$$h_n(x) = O(n^{-\frac{1}{5}}), \quad \|\widehat{f}_{n,h}(x) - f(x)\|_2 = O(n^{-\frac{2}{5}}).$$

For a density of C_s and the L_2 risk, it is $h_n = O(n^{-1/(2s+1)})$ and

$$\|\widehat{f}_{n,h}(x) - f(x)\|_2 = O(n^{-s/(2s+1)}).$$

The bandwidth and the risk decrease as the order of derivability of the density increases.

The derivability condition $f \in C_s$ in 2.1 can be replaced by the condition: f belongs to a Hölder class $\mathcal{H}_{\alpha,M}$ with $|f^{(s)}(x) - f^{(s)}(y)| \leq M|x-y|^{\alpha - s}$ where $s = [\alpha] \geq 0$ is the integer part of $\alpha > 0$.

Proposition 2.4. *Assume f is bounded and belongs to a Hölder class $\mathcal{H}_{\alpha,M}$, then the bias of $\widehat{f}_{n,h}$ is bounded by $Mm_{[\alpha]K}h^\alpha/([\alpha]!) + o(h^\alpha)$, the optimal bandwidth is $O(n^{1/(2\alpha+1)})$ and the MISE at the optimal bandwidth is $O(n^{\alpha/(2\alpha+1)})$.*

2.3 Weak convergence

The L_p-norm of the variations of the process $\widehat{f}_{n,h} - f_{n,h}$ are bounded by the same arguments as the bias and the variance. Assume that K has the support $[-1, 1]$.

Lemma 2.2. *Under Conditions 2.1 and 2.1, there exists a constant C such that for every x and y in $\mathcal{I}_{X,h}$ and satisfying $|x - y| \leq 2h$*

$$E\{\widehat{f}_{n,h}(x) - \widehat{f}_{n,h}(y)\}^2 \leq C(nh^3)^{-1}|x - y|^2.$$

Proof. Let x and y in $\mathcal{I}_{X,h}$, the variance of $\widehat{f}_{n,h}(x) - \widehat{f}_{n,h}(y)$ develops according to their variances given by Proposition 2.2 and the covariance between both terms which has the same bound by the Cauchy-Schwarz inequality. The second order moment $E|\widehat{f}_{n,h}(x) - \widehat{f}_{n,h}(y)|^2$ develops as the sum $n^{-1}\int\{K_{h_n}(x-u) - K_{h_n}(y-u)\}^2 f(u)\,du + (1-n^{-1})\{f_{n,h_n}(x) - f_{n,h_n}(y)\}^2$. For an approximation of the integral $I_2(x,y) = \int\{K_{h_n}(x-u) - K_{h_n}(y-u)\}^2 f(u)\,du$, the Mean Value Theorem implies $K_{h_n}(x-u) - K_{h_n}(y-u) = (x-y)\varphi_n^{(1)}(z-u)$ where $\varphi_n(x) = K_{h_n}(x)$, and z is between x and y, then $\int\{K_{h_n}(x-u) - K_{h_n}(y-u)\}^2 f(u)\,du$ is approximated by

$$(x-y)^2\int\varphi_n^{(1)2}(z-u)f(u)\,du = (x-y)^2 h_n^{-3}\{f(x)\int K^{(1)2} + o(h_n)\}.$$

Since $h_n^{-1}|x|$ and $h_n^{-1}|y|$ are bounded by 1, the order of the second moment of $\widehat{f}_{n,h}(x) - \widehat{f}_{n,h}(y)$ is a $O((x-y)^2(nh_n^3)^{-1})$ if $|x-y| \le 2h_n$ and the covariance is zero otherwise. □

Theorem 2.1. *Under Conditions 2.1 and 2.2, for a density f of class $C_s(\mathcal{I}_X)$ and with nh^{2s+1} converging to a constant γ, the process*

$$U_{n,h} = (nh)^{1/2}\{\widehat{f}_{n,h} - f\}I\{\mathcal{I}_{X,h}\}$$

converges weakly to $W_f + \gamma^{1/2}b_f$, where W_f is a continuous Gaussian process on $\mathcal{I}_X$ with mean zero and covariance $E\{W_f(x)W_f(x')\} = \delta_{x,x'}\sigma_f^2(x)$, at x and x'.

Proof. The finite dimensional distributions of the process $U_{n,h}$ converge weakly to those of $W_f + \gamma^{1/2}b_f$, as a consequence of Proposition 2.2. The covariance of W_f at x and x' is $C_{f,n}(x,x') = \lim_n nhCov\{\widehat{f}_{n,h}(x), \widehat{f}_{n,h}(x')\}$, and Proposition 2.2 implies that $U_{n,h}(x)$ and $U_{n,h}(x')$ are asymptotically independent as n tends to infinity.

If the support of X is bounded, let $a = \inf\mathcal{I}_X$, $\eta > 0$ and $c > \gamma^{1/2}|b_f(a)| + \big(2\eta^{-1}\sigma_f^2(a)\big)^{1/2}$, then

$$\begin{aligned}\Pr\{|U_{n,h}(a)| > c\} &\le \Pr\{(nh)^{1/2}|(\widehat{f}_{n,h} - f_{n,h})(a)| + (nh)^{1/2}|b_{n,h}(a)| > c\}\\ &\le \frac{Var\{(nh)^{1/2}(\widehat{f}_{n,h} - f_{n,h})(a)\}}{\{c - (nh)^{1/2}|b_{n,h}(a)|\}^2},\end{aligned}$$

so that for n sufficiently large

$$\Pr\{|U_{n,h}(a)| > c\} \le \frac{\sigma_f^2(a)}{\{c - \gamma^{1/2}|b_f(a)|\}^2} + o(1) < \eta,$$

the process $U_{n,h}(a)$ is therefore tight. Lemma 2.2 and the bound $\{f_{n,h}(x) - f(x) - f_{n,h}(y) + f(y)\}^2 \le |f(x) - f(y)|^2 + [\int K(z)\{f(x+hz) - f(y+$

$hz)\}dz]^2 \le 2|x-y|^2\|f^{(1)}\|_\infty^2$ imply that the mean of the squared variations of the process $U_{n,h}$ are $O(h^{-2}|x-y|^2)$ as $|x-y| \le 2h < 1$, otherwise the estimators $\widehat{f}_{nh}(x)$ and $\widehat{f}_{nh}(y)$ are independent. Billingsley's Theorem 3 implies the tightness of the process $U_{n,h}1_{[-h,h]}$ and the convergence is extended to any compact subinterval of the support. With an unbounded support for X such that $E|X| < \infty$, for every $\eta > 0$ there exists A such that $P(|X| > A) \le \eta$, therefore $P(|U_{n,h}(A+1)| > 0) \le \eta$ and the same result still holds on $[-A-1, A+1]$ instead of the support of the process $U_{n,h}$. □

Corollary 2.1. *The process*

$$\sup_{x \in \mathcal{I}_{X,h}} \sigma_f^{-1}(x)|U_{n,h}(x) - \gamma^{1/2}b_f(x)|$$

converges weakly to $\sup_{\mathcal{I}_X}|W_1|$, *where* W_1 *is the Gaussian process with mean zero, variance 1 and covariances zero.*

For every $\eta > 0$, *there exists a constant* $c_\eta > 0$ *such that*

$$\Pr\{\sup_{\mathcal{I}_{X,h}} |\sigma_f^{-1}(U_{n,h} - \gamma^{1/2}b_f) - W_1| > c_\eta\}$$

tends to zero as n *tends to infinity.*

Lemma 2.2 concerning second moments does not depend on the smoothness of the density and it is not modified by the condition of a Hölder class instead of a class C_s. The variations of the bias are now bounded by $\{f_{n,h}(x)-f(x)-f_{n,h}(y)+f(y)\}^2 \le 2M\,|x-y|^{2\alpha}$ and the mean of the squared variations of the process $U_{n,h}$ are $O(h^{-2}|x-y|^2)$ for $|x-y| \le 2h < 1$. The weak convergence of Theorem 2.1 is therefore fulfilled with every $\alpha > 1$.

With the optimal bandwidth for the global MISE error

$$h_{AMISE} = \{\frac{\kappa_2}{n m_{2K}^2 \int f^{(2)2}(x)\,dx}\}^{1/5},$$

the limit γ of nh_n^5 is $\kappa_2\, m_{2K}^{-2}\{\int f^{(2)2}(x)\,dx\}^{-1}$. The integral of the second derivative $\int f^{(2)2}(x)\,dx$ and the bias term $b_f = \frac{1}{2}m_{2K}f^{(2)}$ are estimated using the second derivative of the estimator for f. Furthermore, the variance $\sigma_f^2 = \kappa_2 f$ is immediatly estimated. More simply, the asymptotic criterion is written

$$AMISE_n(h) = \int \{h^4 b_f(x) + (nh)^{-1}\sigma_f^2(x)\} f^{-1}(x)\,dF(x)$$

and it is estimated by the empirical mean

$$n^{-1}\sum_{i=1}^{n}\{h^4 b_f(X_i) + (nh)^{-1}\sigma_f^2(X_i)\}f^{-1}(X_i).$$

This empirical error is estimated by

$$\widehat{AMISE}_n(h) = n^{-1}\sum_{i=1}^{n}\{h^4\widehat{b}_{f,n,h_2}(X_i) + (nh)^{-1}\widehat{\sigma}^2_{f,n,h_2}(X_i)\}\widehat{f}_{h_2}^{-1}(X_i)$$

with another bandwidth h_2 converging to zero. The global bandwidth h_{AMISE} is then estimated at the value that achieves the minimum of $\widehat{AMISE}_n(h)$, *i.e.*

$$\widehat{h}_n = \Big\{\frac{4n\sum_{i=1}^{n}\widehat{b}^2_{f,n,h_2}(X_i)\widehat{f}_{h_2}^{-1}(X_i)}{\sum_{i=1}^{n}\widehat{\sigma}^2_{f,n,h_2}(X_i)\widehat{f}_{h_2}^{-1}(X_i)}\Big\}^{-1/5}.$$

Bootstrap estimators for the bias and the variance provide another estimation of $MISE_n(h)$ and h_{AMISE}. These consistent estimators are then used for centering and normalizing the process $\widehat{f}_{n,h} - f$ and provide an estimated process

$$\widehat{U}_n = (n\widehat{h}_n)^{1/2}\widehat{\sigma}^{-1}_{f,n,\widehat{h}_n}\{\widehat{f}_{n,\widehat{h}_n} - f - \widehat{\gamma}_{n,\widehat{h}_n}\widehat{b}_{f,n,\widehat{h}_n}\}I\{\mathcal{I}_{X,\widehat{h}_n}\}.$$

An uniform confidence interval with a level α for the density f is deduced from Corollary 2.1, using a quantile of $\sup_{\mathcal{I}_X}|W_1|$.

Theorem 2.2. *Under Conditions 2.1 and 2.2, for a density f of class $C_s(\mathcal{I}_X)$ and with $nh^{2s+2k+1}$ converging to a constant γ, the process*

$$U^{(k)}_{n,h} = (nh^{2k+1})^{1/2}\{\widehat{f}^{(k)}_{n,h} - f^{(k)}\}I\{\mathcal{I}_{X,h}\}$$

converges weakly to a Gaussian process $W_{f,k} + \gamma^{1/2}b_{f,k}$, where $W_{f,k}$ is a continuous Gaussian process on $\mathcal{I}_X$ with mean and covariances zero.

Let X be a vector variable defined in a subset $\mathcal{I}_X$ of $\mathbb{R}^d$, its density f is estimated by smoothing its distribution function

$$\widehat{F}_n(x) = \sum_{i=1}^{n} 1_{\{X_1\le x_1,\ldots,X_d\le x_d\}},\ x = (x_1,\ldots,x_d),$$

by a multivariate kernel K defined on $[-1,1]^d$ and $K_h(x) = h^{-d}K(h^{-d}x)$, with a single bandwidth or $K_h(x) = \prod_{k=1}^{d} h_k^{-1}K(h_k^{-1}x_k)$ with a vector bandwidth, for x in $\mathcal{I}_{X,h}$. The derivatives of the density $f^{(k)}$ are arrays and the rates of their moments depend on the dimension d. If $h_k = h$

$$\begin{aligned}
b_{s,n,h}(x) &= \frac{h^s}{s!} m_{sK} f^{(s)}(x) + o(h^s), \\
Var\{\widehat{f}_{n,h}(x)\} &= (nh^d)^{-1}\kappa_2 f(x) + o((nh^d)^{-1}), \\
\|\widehat{f}_{n,h}(x) - f_{n,h}(x)\|_p &= 0((nh^d)^{-1/p}), \qquad (2.5) \\
MISE_n(h,x) &= O(h^{2s}) + O((nh^d)^{-1}).
\end{aligned}$$

The optimal bandwidth $h_n(x)$ minimizing the $MISE_n(h,x)$ has the order $n^{-1/(2s+d)}$ where the local MISE reaches the minimal order $O(n^{-2s/(2s+d)})$. The convergence rate of $\widehat{f}_{n,h} - f$ is $(nh^d)^{1/2}$ and the results of Theorem 2.1 and its corollary still hold with this rate.

2.4 Minimax and histogram estimators

Consider a class $\mathcal{F}$ of densities and a risk $R(f, \widehat{f}_n)$ for the estimation of a density f of $\mathcal{F}$ by an estimator $\widehat{f}_n$ belonging to a space $\widehat{\mathcal{F}}$. A minimax estimator $\widehat{f}_n^*$ is defined as a minimizer of the maximal risk over $\mathcal{F}$

$$\widehat{f}_n^* = \arg \inf_{\widehat{f}_n \in \widehat{\mathcal{F}}} \sup_{f \in \mathcal{F}} R(f, \widehat{f}_n).$$

With an optimal bandwidth related to the risk R_p^p, the kernel estimator of a density of $\mathcal{F} = C_s$, $s \geq 2$, provides a L_p-risk of order $h_n^{sp}(x; s, p)$ and this is the minimax risk order in a space $\widehat{\mathcal{F}}$ determined by the regularity of the kernel, the kernel estimator reaches this bound.

The estimator (2.4) of the integral $\theta_k = \int f^{(k)2}(x)\, dx$ of the quadratic k-th derivative of a density of C_2 has therefore the optimal rate of convergence for an estimator of θ_k.

The histogram is the older unsmoothed nonparametric estimator of the density. It is defined as the empirical distribution of the observations cumulated on small intervals of equal length h_n, divided by h_n, with h_n and nh_n converging to zero as n tends to infinity. Let $(B_{jh})_{j=1,\ldots,J_{X,h}}$ be a partition of $\mathcal{I}_X$ into subintervals of length h and centered at a_{jh}, and let $K_h(x) = h^{-1}\sum_{j \in J_{X,h}} 1_{B_{jh}}(x)$ be the kernel corresponding to the histogram, it is therefore defined as

$$\widetilde{f}_{n,h}(x) = hK_h(x) \int K_h(s) d\widehat{F}_n(s).$$

Its bias $\widetilde{b}_{f,h}(x) = \sum_{j \in J_{X,h}} 1_{B_{jh}}(x)\{f(a_{jh}) - f(x)\} + o(h) = hf^{(1)}(x) + o(h)$ is larger than the bias of kernel estimators and its variance $\widetilde{v}_f(x)$ is a

$O((nh)^{-1})$, due to the covariance zero between the empirical distribution on B_{jh} and $B_{j'h}$ for $j \neq j'$. As n tends to infinity, $h^{-1}n\int_{B_{jh}} dVar(\widehat{F}_n - F) = f(a_{jh})\{1 - 2F(a_{jh})\} + o(1)$ hence $\widetilde{v}_{f,h}(x) = (nh)^{-1}f(x)\{1 - 2F(x)\} + o((nh)^{-1})$. Let $\widetilde{b}_f(x) = f^{(1)}(x)$ and $\widetilde{v}_f(x) = f(x)\{1-2F(x)\}$. The normalized histogram $(nh)^{1/2}(\widetilde{f}_{n,h} - f - hf^{(1)})(x)$ converges weakly to a normal variable $\widetilde{v}_{f,h}(x)\mathcal{N}(0,1)$ and it is asymptotically unbiased with a bandwidth $h_n = o(n^{-1/3})$. Increasing the order of h_n reduces the variance of the histogram and increases its bias. The asymptotic mean squared error of the histogram is minimal for the bandwidth

$$h_n(x) = n^{-1/3}\{2\widetilde{b}_f^2(x)\}^{-1/3}\widetilde{v}_f^{1/3}(x) = \{2nf^{(1)2}(x)f^{-1}(x)\}^{-1/3}$$

then it is approximated by

$$\begin{aligned} MSE_{opt}(x) &= n^{-2/3}\{\widetilde{v}_f(x)\widetilde{b}_f(x)\}^{2/3}\{2^{1/3} + 2^{-1/3}\widetilde{b}_f^{2/3}(x)\} \\ &= n^{-2/3}\{f(x)\{1-2F(x)\}f^{(1)}(x)\}^{2/3} \\ &\qquad \times[2^{1/3} + 2^{-1/3}\{f^{(1)}(x)\}^{2/3}]. \end{aligned}$$

These expressions do not depend on the degree of derivability of the density. The optimal bandwidth, the bias $\widetilde{b}_f(x)$ and the variance $\widetilde{v}_f(x)$ of the histogram are estimated by plugging the estimators of the density and its derivative in their formulae. The L_p moments of the histogram are determined by the higher order term in the expansion of $|\widetilde{f}_{n,h}(x) - f_{n,h}(x)|_p^p$, it is a $O((nh)^{-1})$, for every $p \geq 2$.

The derivatives of the density are defined by differences of values of the histogram. For x in B_{jh}, $f^{(1)}(x) = h^{-1}\{f(a_{j+1,h}) - f(a_{j,h})\} + o(1)$ is estimated by

$$\widetilde{f}_{n,h}^{(1)}(x) = h^{-1}\{\widetilde{f}_{n,h}(a_{j+1,h}) - \widetilde{f}_{n,h}(a_{j,h})\}$$

and the derivatives of higher order are defined in the same way. The bias of $\widetilde{f}_{n,h}^{(1)}$ is a $O(1)$ and its variance is a $O((nh^3)^{-1})$.

2.5 Estimation of functionals of a density

The estimation of the integral of a squared density

$$\theta = \int f^2(x)\,dx = \int f(x)\,dF(x)$$

has been considered by many authors and several estimators have been proposed. The plug-in kernel density estimator

$$\widehat{\theta}_{n,h} = \frac{2}{n(n-1)} \sum_{1\leq i<j\leq n} K_h(X_i - X_j)$$

has been introduced by Hall and Marron (1987). A second plug-in estimator was defined by Bickel and Ritov (1988) as the integral of the square of the estimated density

$$\bar{\theta}_{n,h} = \frac{2}{n(n-1)} \sum_{1 \le i < j \le n} \int K_h(x - X_i) K_h(x - X_j)\, dx.$$

Other authors proposed estimators based on projections on orthogonal basis of $L_2(\mathcal{X})$. Bickel and Ritov (1988) and Giné and Nickl (2008) proved the asymptotic equivalence of the estimators and their weak convergence for bounded densities of the space $\mathcal{H}_2^\alpha$ of functions satisfying the integrated Hölder condition $\int \int |t|^{-(1+2\alpha)} |f(x-t) - f(x)|^2 \, dx\, dt < \infty$ for some $0 < \alpha \le 1/2$. Then the bias of the estimator $\widehat{\theta}_{n,h}$ is $(h^{2\alpha})$ and its variance is $(n^2 h \vee n^{-1} h^{2\alpha})$. Moreover, if $\alpha > 1/4$ and $h_n = O(n^{-2/(4\alpha+1)})$, $n^{1/2}\{2\widehat{\theta}_{n,h} - \bar{\theta}_{n,h} - \int f^2(x)\, dx\}$ converges weakly to a centered Gaussian variable with variance $4\{\int f^3 - (\int f^2)^2\}$.

The integrals of the squared k-th derivatives of the density

$$\theta_k = \int f^{(k)2}(x)\, dx$$

are also estimated by the integral of the square of the kernel estimator for the derivative of the density

$$\bar{\theta}_{n,h} = \frac{2}{n(n-1)} \sum_{1 \le i < j \le n} \int K_h^{(k)}(x - X_i) K_h^{(k)}(x - X_j)\, dx,$$

which is equivalent to the integral of the squared estimator (2.3).

The mode of a real function f on an open set $\mathcal{I}_X$ is

$$M_f = \sup_{\mathcal{I}_X} |f(x)|. \tag{2.6}$$

It is a norm on the space of real functions defined on $\mathcal{I}_X$, and for functions with values in a metric space with the norm $|\cdot|$. The triangular inequality provides a bound for the mode of mixture density $g = \sum_{k=1}^p \mu_k f_k$, where the sum of mixture probabilities μ_k belonging to $]0,1[$ satisfies $\sum_{k=1}^p \mu_k = 1$ and the densities f_k have the same support

$$M_g \le \sum_{k=1}^p \mu_k M_{f_k}. \tag{2.7}$$

If the supports of the densities f_k are sufficiently separated, their modes can be identified and the mode of g is identical to the mode of one density of the mixture. Otherwise mixture densities with overlapping supports are cumulated in g and the mode of g is not always located at the mode of a component. It is the combination of the modes of the component densities of the mixture only if all modes have the same location.

For a single density f in class $C_2(\mathcal{I}_X)$, the mode M_f is estimated by

$$\widehat{M}_{f,n,h} = M_{\widehat{f}_{n,h}}.$$

Under Conditions 2.1, the density is locally concave in a neighborhood $\mathcal{N}_M$ of the mode and its estimator has the same property for n sufficiently large. It follows that $\widehat{M}_{f,n,h}$ converges to M_f in probability. A Taylor expansion for x in $\mathcal{N}_M$ is written $f^{(1)}(x) = (x - M_f)f^{(2)}(M_f) + o(x - M_f)$. At the estimated mode, $f^{(1)}(M_f) = 0$ and $\widehat{f}_{n,h}(\widehat{M}_{f,n,h}) = 0$, which entails

$$\begin{aligned}(\widehat{M}_{f,n,h} - M_f) &= f^{(2)-1}(M_f)\, f^{(1)}(\widehat{M}_{f,n,h})\{1 + o(1)\}\\ &= f^{(2)-1}(M_f)\,\{f^{(1)}(\widehat{M}_{f,n,h}) - \widehat{f}^{(1)}_{n,h}(\widehat{M}_{f,n,h})\}\{1 + o(1)\}.\end{aligned}$$

For every x in $\mathcal{I}_X$, the variable $U_{(1),n,h}(x) = (nh^3)^{1/2}(\widehat{f}^{(1)}_{n,h} - f^{(1)})(x)$ converges weakly to a Gaussian variable with a non degenerated variance $\kappa_2 f(x)$ and a mean $m_{(1)}(x) = \lim_n (nh_n^7)^{1/2} m_{2K} f^{(3)}(x)/2$. It follows that the convergence rate of $(\widehat{M}_{f,n,h} - M_f)$ is $(nh^3)^{1/2}$ and

$$(nh^3)^{1/2}(\widehat{M}_{f,n,h} - M_f) = -f^{(2)-1}(M_f)\, U_{(1),n,h}(\widehat{M}_{f,n,h}) + o(1).$$

The following proposition is a consequence of the asymptotic behaviour of the derivative of the estimated density given in Theorem 2.3, the rate of the optimal bandwidth for the density is lower than the rate assumed in Parzen (1962) and the limiting distribution has a non zero mean.

Let $\sigma^2_{M_f} = f(M_f)\{f^{(2)}(M_f)\}^{-2} \int K^{(1)2}(z)\, dz$.

Proposition 2.5. *Under Conditions 2.1 and 2.2, $(nh^3)^{1/2}(\widehat{M}_{f,n,h} - M_f)$ converges weakly to a variable $\mathcal{N}(f^{(2)-1}(M_f)m_{(1)}(M_f), \sigma^2_{M_f})$.*

Note that if $h_n = O(n^{-1/9})$, the convergence rate of $\widehat{M}_{f,n,h}$ is $n^{1/3}$ like Chernoff's estimator (1964). The optimal rate for the bandwidth of the first derivative of the density is $n^{1/(2s+3)}$ as established in Proposition 2.3 for every integer $s > 1$, that is $n^{1/9}$ for a density in $C_3(\mathcal{I}_X)$. For a smaller bandwidth of order $n^{-1/r}$, $r > 2$, $\widehat{M}_{f,n,h}$ converges fastly, with the rate $n^{(1-3/r)/2}$. If the density belongs to $C_3(\mathcal{I}_X)$, the bias of $f^{(1)}(\widehat{M}_{f,n,h})$ is deduced from the bias of the process $(\widehat{f}^{(1)}_{n,h} - f^{(1)})$ and it equals

$$Ef^{(1)}(\widehat{M}_{f,n,h}) = -\frac{h^2}{2}m_{2K}f^{(3)}(\widehat{M}_{f,n,h})+o(h^2) = -\frac{h^2}{2}m_{2K}f^{(3)}(M_f)+o(h^2),$$

it does not depend on the degree of derivability of the density f.

The support of a density f can be estimated from its graph defined as $\mathcal{G}_f = \{(x,y); y = f(x), x \in \mathcal{I}_X\}$. For a continuous function f defined on an open interval $\mathcal{I}_X$ with compact closure, $\mathcal{G}_f$ is an open set of $\mathbb{R}^2$ with compact closure. This closed set defines the support of the function f. For every y such that (x,y) belongs to a closed subset A of $\mathcal{G}_f$, there exist x in a closed subinterval of $\mathcal{I}_X$ such that $y = f(x)$. The graph of a sum of two densities f_1 and f_2 is the union of their graphs $\mathcal{G}_1 \cup \mathcal{G}_2$ and by difference $\mathcal{G}_1 = \mathcal{G}_1 \cup \mathcal{G}_2 - \mathcal{G}_2 \setminus \mathcal{G}_1$, with $\mathcal{G}_2 \setminus \mathcal{G}_1 = \{(x,y); y = f_2(x) \neq f_1(x), x \in \mathcal{I}_X\}$.

Let $\widehat{\mathcal{G}}_{f,n,h} = \{(x,y); y = \widehat{f}_{n,h}(x), x \in \mathcal{I}_X\}$ be the graph of the kernel estimator of an absolutely continuous density f on $\mathcal{I}_X$, then

$$\mathcal{G}_{\widehat{f}_{n,h}} = \mathcal{G}_f \cup \mathcal{G}_{\widehat{f}_{n,h}-f} = \mathcal{G}_f + \mathcal{G}_{\widehat{f}_{n,h}-f} \setminus \mathcal{G}_f$$

hence

$$\mathcal{G}_{\widehat{f}_{n,h}-f} = \widehat{\mathcal{G}}_{f,n,h} - \mathcal{G}_f$$

and it converges a.s. to zero as n tends to infinity. The support of the density f is consistently estimated by $\widehat{\mathcal{G}}_{f,n,h}$ and the extrema of the density are consistently estimated by those of the estimated graph.

2.6 Density of absolutely continuous distributions

Let F_0 be a distribution function in a functional space $\mathcal{F}$ and F_{φ_0} be a distribution function absolutely continuous with respect to F_0, having a density φ_0 with respect to F_0. The function φ_0 belongs to a nonparametric space of continuous functions Φ and the distribution function F_{φ_0} belongs to the nonparametric model $\mathcal{P}_{\mathcal{F},\Phi} = \{(F,\varphi); F \in \mathcal{F}, \varphi \in \Phi, \int_0^\infty \varphi\, dF = 1\}$. The observations are two subsamples $X_1, \ldots, X_{n_1}$ with distribution function F_0 and $X_{n_1+1}, \ldots, X_n$ with distribution function F_{φ_0}. The approach extends straightforwardly to a population stratified in K subpopulations. Estimation of the distributions of stratified populations has already been studied, in particular by Anderson (1979) with a specific parametric form for φ_θ, by Gill, Vardi and Wellner (1988) in biased sampling models with group distributions $\int_0^\cdot w_k\, dF$, where the weight functions are known, by Gilbert (2000) in biased sampling models with parametric weight functions, by Cheng and Chu (2004), with the Lebesgue measure and kernel density estimators.

The density with respect to the Lebesgue measure of a distribution function F in $\mathcal{F}$ is denoted f and the distributions of both samples are supposed to have the same support. Let $n_2 = n - n_1$ increasing with n, such that $\lim_n n^{-1}n_1 = \pi$ in $]0,1[$, and let ρ be the sample indicator defined by $\rho = 1$ for individuals of the first sample and $\rho = 0$ for individuals of the second sample. Let $F_1 = \pi F_0$ and $F_2 = (1-\pi)F_{\varphi_0}$ be the subdistribution functions of the two subsamples, they are estimated by the corresponding empirical subdistribution functions

$$\widehat{F}_{1,n} = n^{-1}\sum_{i=1}^{n}\rho_i 1_{\{X_i \le t\}} = n^{-1}\sum_{i=1}^{n_1} 1_{\{X_i^1 \le t\}},$$
$$\widehat{F}_{2,n} = n^{-1}\sum_{i=1}^{n}(1-\rho_i)1_{\{X_i \le t\}}$$

and $\widehat{\pi}_n = n^{-1}n_1$. Their densities with respect to the Lebesgue measure are denoted f_1 and f_2, and the density of the second sample with respect to the distribution of the first one is $\varphi = \pi(1-\pi)^{-1}f_1^{-1}f_2$. The densities f_1 and f_2 are estimated by smoothing $\widehat{F}_{1,n}$ and $\widehat{F}_{2,n}$, then f_0, f_φ and φ are estimated by

$$\widehat{f}_{0,n,h}(t) = \widehat{\pi}_n^{-1}\int K_h(t-s)\,d\widehat{F}_{1,n}(s),$$
$$\widehat{f}_{n,h}(t) = (1-\widehat{\pi}_n)^{-1}\int K_h(t-s)\,d\widehat{F}_{2,n}(s),$$
$$\widehat{\varphi}_{n,h_n}(t) = \widehat{f}_{0,n,h}^{-1}(t)\widehat{f}_{n,h}(t)$$

on every compact subset of the support of the densities where f_0 is strictly positive and $\|\widehat{\varphi}_{n,h} - \varphi_0\|$ converges in probability to 0. The expectation of the estimators are approximated by $f_{0;n,h}(t) = \int K_h(t-s)\,dF_0(s) + O(n^{-1/2}) = f_0 + \frac{h^2}{2}f_0^{(2)} + o(h^2)$ and $f_{n,h}(t) = \int K_h(t-s)\,dF_\varphi(s) + O(n^{-1/2}) = f_\varphi + \frac{h^2}{2}f_\varphi^{(2)} + o(h^2)$. The bias of $\widehat{\varphi}_{n,h}$ is expanded as $b_{n,h} = \frac{h^2}{2}f_0^{-1}\{\varphi f_0^{(2)} + 2\varphi^{(1)}f_0^{(1)}\}m_{2K} + o(h^2)$, its variance is $v_{n,h} = f_0^{-2}\{Var\widehat{f}_{n,h} + \varphi^2\, Var\widehat{f}_{0,n,h}\} + o((nh)^{-1})$ where the variances given in Proposition 2.2, $Var\widehat{f}_{j,n,h}(t) = (n_jh)^{-1}\kappa_2 f_j(t)\{1+o(1)\}$ imply similar approximations for the variances of the estimators $\widehat{f}_{0,n,h}$ and $\widehat{f}_{n,h}$. The following approximation with independent subsamples implies the weak convergence of the estimator of φ

$$(nh)^{1/2}(\widehat{\varphi}_{n,h} - \varphi_{n,h}) = f_0^{-1}(nh)^{1/2}\{(\widehat{f}_{n,h} - f_{\varphi,n,h}) - \varphi(\widehat{f}_{0,n,h} - f_{0,n,h})\} + o_{L_2}(1).$$

2.7 Hellinger distance between a density and its estimator

Let P and Q be two probability measures and let $\lambda = P + Q$ be the dominating measure of their sum. Let F and G be the distribution functions of a variable X under the probability measures P and Q, respectively, and let f and g be the densities of P and Q, respectively, with respect to λ. The Hellinger distance between P and Q is

$$h^2(P,Q) = \frac{1}{2}\int (\sqrt{dP} - \sqrt{dQ})^2 = \frac{1}{2}\int (\sqrt{f} - \sqrt{g})^2 d\lambda.$$

The affinity of P and Q is

$$\rho(P,Q) = 1 - h^2(P,Q) = \int \sqrt{fg}\, d\lambda.$$

The following inequalities were proved by Lecam and Yang (1990)

$$h^2(P,Q) \le \frac{1}{2}\|P - Q\|_1 \le \{1 - \rho^2(P,Q)\}^{1/2}.$$

Applying this inequality to the probability density f of P, absolutely continuous with respect to the Lebesgues measure λ, and its estimator $\widehat{f}_{n,h}$, we obtain

$$h^2(\widehat{f}_{n,h}, f) = \int (1 - \sqrt{\frac{\widehat{f}_{n,h}}{f}})\, dF \le \{1 - (\int \sqrt{\frac{\widehat{f}_{n,h}}{f}}\, dF)^2\}^{1/2}.$$

The convergence to zero of the Hellinger distance $h^2(\widehat{f}_{n,h}, f)$ is deduced from the obvious bound

$$h^2(\widehat{f}_{n,h}, f) = \int (1 - \sqrt{\frac{\widehat{f}_{n,h}}{f}})\, dF \le \int (\sqrt{\frac{\widehat{f}_{n,h}}{f}} - 1)\, d(\widehat{F}_n - F) \qquad (2.8)$$

which is consequence of the inequality $\int \sqrt{\frac{\widehat{f}_{n,h}}{f}}\, d\widehat{F}_n \ge 0$. This inequality and the uniform *a.s.* consistency of the density estimator also imply the *a.s.* convergence to zero of $n^{1/2}h^2(\widehat{f}_{n,h}, f)$. By differentiation, estimators of functionals of the density converges with the same rate as the estimator of the density, hence $h^2(\widehat{f}_{n,h}, f)$ convergences to zero in probability with the rate $nh_n^{1/2}$.

Applying these results to the probability measures P_0 and $P = P_{\varphi_0}$ of the previous section, with distribution functions F_0 and F, we get similar formulae

$$h^2(P_0, P) = \frac{1}{2}\int (1 - \sqrt{\varphi})^2 dF_0 \le \{1 - (\int \sqrt{\varphi}\, dF_0)^2\}^{1/2},$$

$$h^2(\widehat{\varphi}_{n,h}, \varphi) = \int (1 - \sqrt{\frac{\widehat{\varphi}_{n,h}}{\varphi}})\, dF \le \{1 - (\int \sqrt{\frac{\widehat{\varphi}_{n,h}}{\varphi}}\, dF)^2\}^{1/2}.$$

The bound (2.8) is adapted to the density φ

$$h^2(\widehat{\varphi}_{n,h},\varphi) \le \int \Big(\sqrt{\frac{\widehat{\varphi}_{n,h}}{\varphi}} - 1\Big)\, d(\widehat{F}_n - F),$$

it follows that the convergence rate of $h^2(\widehat{\varphi}_{n,h_n},\varphi)$ is $nh_n^{1/2}$.

2.8 Estimation of the density under right-censoring

On a probability space $(\Omega, \mathcal{A}, P)$, let X and C be two independent positive random variables with densities f and f_C and such that $P(X < C)$ is strictly positive, and let $T = X \wedge C$, $\delta = 1_{\{X \le C\}}$ denote the observed variables when X is right-censored by C. Let

$$N_n(t) = \sum_{1 \le i \le n} \delta_i 1_{\{T_i \le t\}}$$

be the number of observations before t and

$$Y_n(t) = \sum_{1 \le i \le n} 1_{\{T_i \ge t\}}$$

be the number of individuals at risk at t. The survival function $\bar{F} = 1 - F^-$ of X is now estimated by Kaplan-Meier's product-limit estimator

$$\widehat{\bar{F}}{}_n^R(t) = \prod_{T_i \le t} \{1 - \frac{\delta_i J_n(T_i)}{Y_n(T_i)}\} = \prod_{T_i \le t} \{1 - \Delta\widehat{\Lambda}_n(T_i)\}, \text{ with}$$

$$\widehat{\Lambda}_n(t) = \int_0^t \frac{J_n(s)}{Y_n(s)}\, dN_n(s), \tag{2.9}$$

and $J_n(s) = 1_{\{Y_n(s)>0\}}$. The process $\widehat{F}_n^R$ is also written in an additive form (Pons, 2007) as a right-continuous increasing process identical to the product-limit estimator

$$\widehat{F}_n^R(t) = \int_0^t \frac{dN_n(s)}{n - \sum_{j=1}^n (1-\delta_j) 1_{\{T_j < s\}} \{1 - (\widehat{F}_n^R(T_j))^{-1}\}} \tag{2.10}$$

which is easily calculated. From the martingale property of the process $\widehat{\Lambda}_n$ and Gill's expression for the Kaplan-Meier estimator

$$\frac{\widehat{F}_n^R - F}{1 - F}(t) = -\int_0^t \frac{1 - \widehat{F}_n^R(s^-)}{1 - F(s)}\, d(\widehat{\Lambda}_n - \Lambda)(s), \; t \le \max T_i, \tag{2.11}$$

it follows that

$$E \int_0^t \frac{1 - \widehat{F}_n^R(s^-)}{1 - F(s)} \{d\widehat{\Lambda}_n(s) - d\Lambda(s)\} = 0,$$

so the Kaplan-Meier estimator is unbiased and for every $t \leq \max T_i$, its variance is $Var\widehat{F}_n^R(t) = (1-F)^2(t)\int_0^t g^{-1}\,d\Lambda + o(1)$. For every $p \geq 2$, the Burkhölder-Davis-Gundy inequality implies the existence of a constant c_p such that a bound for the L_p-risk of the Kaplan-Meier estimator is written

$$E\{F(t) - \widehat{F}_n^R(t)\}^p \leq c_p\{1-F(t)\}^p E\Big[\int_0^t \{\frac{1-\widehat{F}_n^R(s^-)}{1-F(s)}\}^2 Y_n^{-2}(s)dN_n(s)\Big]^{p/2}$$

therefore $\|F(t) - \widehat{F}_n^R(t)\|_p = O(n^{1/2})$.

The density of T under right-censoring is estimated by smoothing the Kaplan-Meier estimator $\widehat{F}_n^R$ of the distribution function

$$\widehat{f}_{n,h}^R(t) = \int K_h(t-s)\,d\widehat{F}_n^R(s)$$

it is explicitly written using either its multiplicative expression

$$\widehat{f}_{n,h}^R(t) = \int_{\mathcal{I}_{X,n,h}} K_h(t-s)\widehat{F}_n^R(s^-)\,d\widehat{\Lambda}_n(s),$$

or the additive formula (2.10)

$$\widehat{f}_{n,h}^R(t) = \sum_{i=1}^n \frac{K_h(t-T_i)\delta_i 1_{\{T_i \leq t\}}}{n - \sum_{j=1}^n (1-\delta_j)1_{\{T_j < T_i\}}\{1 - (\widehat{F}_n^R(T_j))^{-1}\}}.$$

The *a.s.* uniform consistency of the process $\widehat{F}_n^R - F$, for the Kaplan-Meier estimator, implies that $\sup_{\mathcal{I}_{X,h}} |\widehat{f}_{n,h}^R - f|$ converges in probability to zero, as n tends to the infinity and h to zero. From (2.11), the estimator $\widehat{f}_{n,h}^R$ satisfies

$$\begin{aligned}
\widehat{f}_{n,h}^R(t) &= \int K_h(t-s)[f(s)\{1 + \int_0^s \frac{1-\widehat{F}_n^{R-}}{1-F}\,d(\widehat{\Lambda}_n - \Lambda)\}\,ds \\
&\qquad - \{1 - \widehat{F}_n^R(s^-)\}\,d(\widehat{\Lambda}_n - \Lambda)(s)] \\
&= \int K_h(t-s)\,[dF(s) - \{1 - \widehat{F}_n^R(s^-)\}\,d(\widehat{\Lambda}_n - \Lambda)(s)] \\
&\quad + \int \{\int_h^u K_h(t-s)\,dF(s)\}\frac{1-\widehat{F}_n^{R-}}{1-F}(u)\,d(\widehat{\Lambda}_n - \Lambda)(u).
\end{aligned}$$

The bias of the estimated density $\widehat{f}_{n,h}^R(t)$ is then the same as in the uncensored case $b_{f,n,h}(t) = \frac{h^2}{2}f^{(2)}(t) + o(h^2)$, since the Kaplan-Meier estimator and $\widehat{\Lambda}_n$ are unbiased estimators. Its variance develops as

$$\begin{aligned}
v^R_{f,n,h}(t) &= E\int K^2_h(t-s)f^2(s)\Big\{\int_0^s \frac{1-\widehat{F}^{R-}_n}{1-F}\,d(\widehat{\Lambda}_n-\Lambda)\Big\}^2\,ds \\
&\quad + E\int K^2_h(t-s)\{1-\widehat{F}_n(s^-)\}^2 Y_n^{-1}(s)\,d\Lambda(s) \\
&\quad - 2E\int K^2_h(t-s)\frac{(1-\widehat{F}^{R-}_n)^2}{1-F}(s)Y_n^{-1}(s)\,d\Lambda(s) \\
&\le (nh)^{-1}\kappa_2\{f^2(t)\int_0^t g^{-1}\,d\Lambda + (1-F_C)^{-1}(t)f(t) \\
&\quad - 2f(t)g^{-1}(t) + o(1)\},
\end{aligned}$$

where

$$E\Big\{\int_0^s \frac{1-\widehat{F}^{R-}_n}{1-F}\,d(\widehat{\Lambda}_n-\Lambda)\Big\}\Big\}^2 \le E\int_0^s \Big\{\frac{1-\widehat{F}^{R-}_n}{1-F}\Big\}^2 Y_n^{-1}\,d\Lambda.$$

The variance is then written $v^R_{f,n,h}(t) = (nh)^{-1}v^R_f(t)$. It follows that the optimal local and global bandwidths for the estimation of the density under right-censoring are $O(n^{-1/5})$ and the optimal L_2-risks are $O(n^{-2/5})$.

2.9 Estimation of the density of left-censored variables

Let X be left-censored by C, such that $P(C < X)$ is strictly positive, then the observations are $T = X \vee C$, $\delta = 1_{\{C<X\}}$. The notations N_n and Y_n for a sample of n independent and identically distributed observations of $(T_i,\delta_i)_{1\le i\le n}$ are unchanged, with this definition of the variables. The cumulative hazard function λ used for right-censoring is replaced by a cumulative retro-hazard function (Pons, 2008)

$$\bar{\Lambda}(t) = \int_t^\infty \frac{dF}{F^-}$$

or $\bar{\Lambda}(t) = \int_t^\infty \frac{dF}{F^-} + \sum_{s>t}\frac{\Delta F(s)}{F(s^-)}$ which uniquely defines the distribution function F of X, for $t > \min_i T_i$, as

$$F(t) = \exp\{\bar{\Lambda}^c(t)\}\prod_{s>t}\{1+\Delta\bar{\Lambda}(s)\},$$

where $\bar{\Lambda}^c(t)$ is the continuous part of $\bar{\Lambda}(t)$ and $\prod_{s>t}\{1+\Delta\bar{\Lambda}(s)\}$ its right-continuous discrete part. On the interval $\mathcal{I}_n =]\min_i T_i, \max_i T_i]$, the function $\bar{\Lambda}$ is estimated by

$$\widehat{\bar{\Lambda}}_n(t) = \int_t^\infty 1\{Y_n < n\}\frac{dN_n}{n-Y_n}$$

and a product-limit estimator of the function F is defined on $\mathcal{I}_n$ from the expression of $\widehat{\bar\Lambda}_n$ by

$$\widehat{F}_n^L(t) = \prod_{T_i \geq t} \left\{1 + d\widehat{\bar\Lambda}_n(T_i)\right\}^{\delta_i}.$$

On the interval $\mathcal{I}_n =]\min_i T_i, \max_i T_i]$ it satisfies

$$\frac{F - \widehat{F}_n^L}{F}(t) = \int_t^\infty \frac{\widehat{F}_n^L(s^-)}{F(s)} \{d\widehat{\bar\Lambda}_n(s) - d\bar\Lambda(s)\}, \tag{2.12}$$

and $n^{1/2}(F - \widehat{F}_n^L)F^{-1}$ converges weakly to a centered Gaussian process with covariance $\bar K(s,t) = \int_{s\wedge t}^\infty (F^{-1}F^-)^2 (H^-)^{-1} \, d\bar\Lambda$, with $1 - H = (1-F)(1 - F_C)$. From this expression, it follows that $\widehat{\bar\Lambda}_n$ is an unbiased estimator of $\bar\Lambda$ and $\widehat{F}_n^L$ is an unbiased estimator of the distribution function F, moreover $\|\widehat{F}_n^L(t) - F(t)\|_p = O(n^{1/2})$, for $p \geq 2$.

The density of T under left-censoring is estimated by smoothing the Kaplan-Meier estimator $\widehat{F}_n^L$ of the distribution function

$$\widehat{f}_{n,h}^L(t) = \int K_h(t-s) \, d\widehat{F}_n^L(s).$$

The a.s. uniform consistency of the process $\widehat{F}_n^L - F$ implies that $\sup_{\mathcal{I}_{X,h}} |\widehat{f}_{n,h}^L - f|$ converges in probability to zero, as n tends to the infinity and h to zero. From (2.12), the estimator $\widehat{f}_{n,h}^L$ satisfies

$$\begin{aligned}
\widehat{f}_{n,h}^L(t) &= f_{n,h}(t) + \int K_h(t-s)[f(s)\{\int_s^\infty \frac{\widehat{F}_n^{L-}}{F} \, d(\widehat{\bar\Lambda}_n - \bar\Lambda)\} \, ds \\
&\qquad - \widehat{F}_n^{L-}(s) \, d(\widehat{\bar\Lambda}_n - \bar\Lambda)(s)] \\
&= f_{n,h}(t) + \int_t^\infty f_{n,h}(s) \frac{\widehat{F}_n^L(s^-)}{F(s)} \, d(\widehat{\bar\Lambda}_n - \bar\Lambda)(s) \\
&\qquad - \int K_h(t-s) \widehat{F}_n^L(s^-) \, d(\widehat{\bar\Lambda}_n - \bar\Lambda)(s).
\end{aligned}$$

As a consequence of the uniform consistency of the estimators $\widehat{F}_n^{L-}$ and $\widehat{\bar\Lambda}_n$, the bias of the estimated density $\widehat{f}_{n,h}^L(t)$ is then the same as in the uncensored case $b_{f,n,h}(t) = \frac{h^2}{2} f^{(2)}(t) + o(h^2)$. Its variance is written $v_{f,n,h}^L(t) = (nh)^{-1} v_f^L(t)$, with the expansion

$$\begin{aligned}
v_{f,n,h}^L(t) &= E \int_t^\infty f_{n,h}^2(s) \{\frac{\widehat{F}_n^L(s^-)}{F(s)}\}^2 (n - Y_n(s))^{-1} \, d\bar\Lambda(s) \\
&\quad + \int K_h^2(t-s) \widehat{F}_n^{L2}(s^-)(n - Y_n(s))^{-1} \, d\bar\Lambda(s) \\
&\quad - 2 \int_t^\infty f_{n,h}(s) \frac{\widehat{F}_n^{L2}(s^-)}{F(s)} K_h(t-s)(n - Y_n(s))^{-1} \, d\bar\Lambda(s)
\end{aligned}$$

where the last two terms are $O((nh)^{-1})$ and the first one is a $O(n^{-1})$. The optimal bandwidths for estimating the density under left-censoring are then also $O(n^{-1/5)}$ and the optimal L_2-risks are $O(n^{-2/5})$.

Under Conditions 2.1 or 2.2 and if the support of K is compact, the variance v_f^L belongs to class $C_2(\mathcal{I}_X)$ and for every t and t' in $\mathcal{I}_{X,h}$, there exists a constant α such that for $|t-t'| \leq 2h$

$$E\{\widehat{f}_{n,h}^L(t) - \widehat{f}_{n,h}^L(t')\}^2 \leq \alpha(nh^3)^{-1}|t-t'|^2.$$

Under the conditions of Theorem 2.1, the process $U_{n,h}^L = (nh)^{1/2}\{\widehat{f}_{n,h}^L - f\}I\{\mathcal{I}_{X,h}\}$ converges weakly to $W_f^L + \gamma^{1/2}b_f$, where W_f^L is a continuous Gaussian process on $\mathcal{I}_X$ with mean and covariances zero and with variance function v_f^L.

2.10 Kernel estimator for the density of a process

Consider a continuously observed stationary process $(X_t)_{t\in[0,T]}$ with values in $\mathcal{I}_X$. The stationarity means that the distribution probability of X_t and $X_{t+s} - X_s$ are identical for every s and $t > 0$. For a process with independent increments, this implies the ergodicity of the process that is expressed by the convergence of bounded functions of several observations of the process to a mean value: For every x in $\mathcal{I}_X$, there exists a measure π_x on $\mathcal{I}_X \setminus \{x\}$ such that for every bounded and continuous function ψ on $\mathcal{I}_X^2$

$$ET^{-1}\int_{[0,T]^2} \psi(X_s, X_t)\, ds\, dt \to \int_{\mathcal{I}_X^2} \psi(x,y)\, d\pi_x(dy)dF(x) \tag{2.13}$$

as T tends to infinity. The distribution function F in (2.13) is defined as the limit of the expectation of the mean sample-path of the process X

$$ET^{-1}\int_{[0,T]} \psi(X_t)\, dt \to \int_{\mathcal{I}_X} \psi(x)\, dF(x). \tag{2.14}$$

The mean marginal density f of the process is the density of the distribution function F, it is estimated by replacing the integral of a kernel function with respect to the empirical distribution function of a sample by an integral with respect to the Lebesgue measure over $[0,T]$ and the bandwidth sequence is indexed by T. For every x in $\mathcal{I}_{X,T,h}$

$$\widehat{f}_{T,h}(x) = \frac{1}{T}\int_0^T K_h(X_s - x)\, ds, \tag{2.15}$$

its expectation is $f_{T,h}(x) = \int_{\mathcal{I}_{X,n}} K_h(y-x)f(y)\,dy$ so its bias is

$$b_{T,h}(x) = \int_{\mathcal{I}_{X,T}} K_h(y-x)\{f(y)-f(x)\}\,dy = \frac{h_T^s}{s!} m_{sK} f^{(s)}(x) + o(h_T^s)$$

under Conditions 2.1-2.2. For a density in a Hölder class $\mathcal{H}_{\alpha,M}$, $b_{T,h}(x)$ tends to zero for every $\alpha > 0$ and it is a $O(h^{[\alpha]})$ under the condition $\int |u|^{[\alpha]} K(u)\,du < \infty$.

Its variance is expressed through the integral of the covariance between $K_h(X_s - x)$ and $K_h(X_t - x)$. For $X_s = X_t$, the integral on the diagonal $\mathcal{D}_X$ of $\mathcal{I}_{X,T}^2$ is a $(Th_T)^{-1}\kappa_2 f(x) + o((Th_T)^{-1})$ and the integral outside the diagonal denoted $I_o(T)$ is expanded using the ergodicity property (2.13). Let $\alpha_h(u,v) = |u-v|/2h_T$, the integral $I_o(T)$ is written

$$\begin{aligned}&\int_{[0,T]^2}\int_{\mathcal{I}_{X,T}^2\setminus\mathcal{D}_X} K_h(u-x)K_h(v-x) f_{X_s,X_t}(u,v)\,du\,dv\,\frac{ds}{T}\frac{dt}{T}\\ &= (Th_T)^{-1}\{\int_{\mathcal{I}_X}\int_{\mathcal{I}_{X\setminus\{u\}}}\int_{-1+\alpha_h(u,v)}^{1-\alpha_h(u,v)} K(z-\alpha_h(u,v))K(z+\alpha_h(u,v))\,dz\\ &\qquad d\pi_u(v)\,dF(u)\}\{1+o(1)\}.\end{aligned}$$

For every fixed $u \neq v$, the integral $\int K(z-\alpha_h(u,v))K(z+\alpha_h(u,v))\,dz$ tends to zero since $\alpha_{h_T}(u,v)$ tends to infinity as h_T tends to zero. If $\alpha_h(u,v)$ tends to zero with h_T, then $\pi_u(v)$ also tends to zero and the integral $I_o(T)$ is a $o((Th_T)^{-1})$ as T tends to infinity. The mean squared error of the estimator at x for a marginal density in C_s is then

$$\begin{aligned}MISE_{T,h}(x) &= (Th_T)^{-1}\kappa_2 f(x) + h_T^{2s}(s!)^{-2} m_{sK}^2 f^{(s)2}(x)\\ &\quad + o((Th_T)^{-1}) + o(h_T^{2s})\end{aligned}$$

and the optimal local and global bandwidths minimizing the mean squared (integrated) errors are $O(T^{1/(2s+1)})$. If h_T has the rate of the optimal bandwidths, the $MISE$ is a $O(T^{2s/(2s+1)})$. The L_p-norm of the estimator satisfies $\|\widehat{f}_{T,h}(x) - f_{T,h}(x)\|_p = O((Th_T)^{-1/p})$ under an ergodicity condition for $(X_{t_1},\ldots,X_{t_k})$ similar to (2.13) for bounded functions ψ defined on $\mathcal{I}_X^k$

$$\begin{aligned}&ET^{-1}\int_{[0,T]^k} \psi(X_{t_1},\ldots,X_{t_k})\,dt_1\ldots dt_k \qquad (2.16)\\ &\quad\to \int_{\mathcal{I}_X^k} \psi(x_1,\ldots,x_k) \prod_{1\le j\le k-1} \pi_{x_j}(dx_{j+1})\,dF(x_1),\end{aligned}$$

for every integer $k = 2,\ldots,p$. The property (2.16) implies the weak convergence of the finite dimensional distributions of the process $(Th_T)^{1/2}(\widehat{f}_{T,h} -$

$f - b_{T,h})$ to those of a centered Gaussian process with mean zero, covariances zero and variance $\kappa_2 f(x)$ at x. The proof is similar to the proof for a sample of variables, using the above expansions for the variance and covariances of the process. A lipschitzian bound for increments $E\{\widehat{f}_{T,h}(x) - \widehat{f}_{T,h}(y)\}^2$ is obtained by the Mean Value Theorem which implies $T^{-2}\int_0^T E\{K_{h_n}(x-X_t) - K_{h_n}(y-X_t)\}^2 dt = O(|x-y|^2(Th_T^3)^{-1})$ as in Lemma 2.1. Then the process $(Th_T)^{1/2}(\widehat{f}_{T,h} - f - b_{T,h})$ converges weakly to a centered Gaussian process with covariances zero and variance $\kappa_2 f$.

The Hellinger distance $h^2(\widehat{f}_{T,h_T}, f)$ is bounded like (2.8)

$$h_T^2(\widehat{f}_{T,h_T}, f) = \int (1 - \sqrt{\frac{\widehat{f}_{T,h_T}}{f}})\, dP \le \int (\sqrt{\frac{\widehat{f}_{T,h_T}}{f}} - 1)\, d(\widehat{F}_T - F)$$

where

$$\widehat{F}_T(t) = T^{-1} \int_0^T 1_{\{X_t \le s\}}\, dt$$

is the empirical probability distribution of the mean marginal distribution function of the process $(X_t)_{t \le T}$

$$F_T = T^{-1} \int_{[0,T]} F_{X_t}\, dt,$$

and F is its limit under the ergodicity property (2.14). The convergence rate of $\widehat{F}_T - F$ is $\sqrt{T}$, from the mixing property of the process X. Therefore $h^2(\widehat{f}_{T,h_T}, f)$ convergences to zero in probability with the rate $Th_T^{1/2}$.

2.11 Exercises

(1) Let f and g be real functions defined on $\mathbb{R}$ and let $f * g(x) = \int f(x-y)g(y)\, dy$ be their convolution. Calculate $\int f * g(x)\, dx$ and prove that, for $1 \le p \le \infty$, if f belongs to L_p and g to L_q such that $p^{-1} + q^{-1} = 1$, then $\sup_{x \in \mathbb{R}} |f * g(x)| \le \|f\|_p \|g\|_q$. Assume p is finite and prove that $f * g$ belongs to the space of continuous functions on $\mathbb{R}$ tending to zero at infinity.

(2) Prove the approximation of the bias in (d) of Proposition 2.2 using a Taylor expansion and precise the expansions for the L_p-risk.

(3) Prove the results of Equation (2.5).

(4) Write the variance of the kernel estimator for the marginal density of dependent observations $(X_i)_{i\leq n}$ in terms of the auto-covariance coefficients $\rho_j = n^{-1}\sum_{i=1}^n Cov(X_i, X_{i+j})$.
(5) Consider a hierarchical sample $(X_{ij}, Y_{ij})_{j=(1,\dots,J_i),i=1,\dots,n}$, with n independent and finite sub-samples of J_i dependent observations. Let $N = \sum_{i=1}^n J_i$ and $f = \lim_n N^{-1}\sum_{i=1}^n \sum_{j=1}^{J_i} f_{X_{ij}}$ be the limiting marginal mean density of the observations of X. Define an estimator of the density f and give the first order approximation of the variance of the estimator under relevant ergodicity conditions.
(6) Let $H(x) = \int_{-1}^x k(y)\,dy$ be the integrated kernel, F be the distribution function of X and $\widehat{F}_{nh}(x) = n^{-1}\sum_{i=1}^n H_h(X_i - x)$ be a smooth estimator of the distribution function. Prove that the bias of $\widehat{F}_{nh}(x)$ is $\frac{1}{2}h^2 m_{2K} f^{(1)}(x) + o(h^2)$ and its variance $(nh)^{-1}\kappa_2 F(x) + o((nh)^{-1})$. Define the optimal local and global bandwidths for $\widehat{F}_{nh}$.

Chapter 3

Kernel estimator of a regression function

3.1 Introduction and notation

The kernel estimation of nonparametric regression functions is related to the estimation of the conditional density of a variable and most authors have studied the asymptotic behaviour of weighted risks, using weights proportional to the density estimator so that the random denominator of the regression function disappears. Weighted integrated errors are used for the empirical choice of a bandwidth and for tests about the regression. In this chapter, the bias, variance and norms of the kernel regression estimator are obtained from a linear approximation of the estimator.

Let $(X_i, Y_i)_{i=1,\ldots,n}$ be a sample of a variable (X, Y) with joint density $f_{X,Y}$. The marginal density of X is $f_X(x) = \int f_{X,Y}(x, y)dy$ and the density of Y conditionally on X is $f_{Y|X} = f_X^{-1} f_{X,Y}$. Here, the density $f_{X,Y}$ is supposed to be C_2. Let F_{XY} be the distribution function of (X, Y) and $\widehat{F}_{XY,n}(x, y) = n^{-1} \sum_{i=1}^n 1\{X_i \leq x, Y_i \leq y\}$ be their empirical distribution function.

Consider the regression model (1.6)

$$Y = m(X) + \sigma\varepsilon$$

where m is a bounded function and the error variable ε has the conditional mean $E(\varepsilon|X) = 0$ and a constant conditional variance $Var(\varepsilon|X) = \sigma^2$. Let $\mathcal{I}_X$ and $\mathcal{I}_{XY}$ be respectively subsets of the supports of the distribution functions F_X and F_{XY}, and let

$$\mathcal{I}_{X,h} = \{x \in \mathcal{I}_X; [x - h, x + h] \in \mathcal{I}_X\},$$

$$\mathcal{I}_{XY,h} = \{(x, y) \in \mathcal{I}_{XY}; [x - h, x + h] \times \{y\} \in \mathcal{I}_{XY}\}$$

be subsets of the supports. On an interval $\mathcal{I}_{XY,h}$, a continuous and bounded regression function

$$m(x) = E(Y|X = x) = f_X^{-1} \int y f_{XY}(x, y)\, dy$$

is estimated by the kernel estimator

$$\widehat{m}_{n,h}(x) = \frac{\sum_{i=1}^n Y_i K_h(x - X_i)}{\sum_{i=1}^n K_h(x - X_i)}. \tag{3.1}$$

Its numerator is denoted

$$\widehat{\mu}_{n,h}(x) = \frac{1}{n} \sum_{i=1}^n Y_i K_h(x - X_i) = \int y K_h(x - s)\, d\widehat{F}_{XY,n}(s, y)$$

and its denominator is $\widehat{f}_{X,n,h}(x)$. The mean of $\widehat{\mu}_{n,h}(x)$ and its limit are respectively

$$\mu_{n,h}(x) = \int \int y K_h(x - s)\, dF_{XY}(s, y),$$
$$\mu(x) = \int y f_{XY}(x, y)\, dy = f_X(x) m(x),$$

whereas the mean of $\widehat{m}_{n,h}(x)$ is denoted $m_{n,h}(x)$. The notations for the parameters and estimators of the density f are unchanged. The variance of Y is supposed to be finite and its conditional variance is denoted

$$\sigma^2(x) = E(Y^2|X = x) - m^2(x),$$
$$E(Y^2|X = x) = f_X^{-1}(x) w_2(x) = \int y^2 f_{Y|X}(y; x)\, dy, \text{ with}$$
$$w_2(x) = \int y^2 f_{XY}(x, y)\, dy = f_X(x) \int y^2 f_{Y|X}(y; x)\, dy.$$

Let also $\sigma_4(x) = E[\{Y - m(x)\}^4 \mid X = x]$, they are supposed to be bounded functions. The L_p-risk of the kernel estimator of the regression function m is defined by its L_p-norm $\| \cdot \|_p = \{E\| \cdot \|^p\}^{1/p}$.

3.2 Risks and convergence rates for the estimator

The following conditions are assumed, in addition to Conditions 2.1 and 2.2 about the kernel and the density.

Condition 3.1. (1). The functions f_X, m and μ are twice continuously differentiable on $\mathcal{I}_X$, with bounded second order derivatives; f_X is strictly positive on $\mathcal{I}_X$;

(2). The functions f_X , m and σ belong to the class $C_s(\mathcal{I}_X)$.

Proposition 3.1. *Under Conditions 2.1, 2.2 and 3.1(1),*

(a). $\sup_{x\in\mathcal{I}_{X,h}}|\widehat{\mu}_{n,h}(x)-\mu(x)|$ *and* $\sup_{x\in\mathcal{I}_{X,h}}|\widehat{m}_{n,h}(x)-m(x)|$ *converge a.s. to zero if and only if* μ *and* m *are uniformly continuous.*

(b). The following expansions are satisfied

$$m_{n,h}(x)=\frac{\mu_{n,h}(x)}{f_{X,n,h}(x)}+O((nh)^{-1}),$$

$$\begin{aligned}(nh)^{1/2}\{\widehat{m}_{n,h}-m_{n,h}\}(x)&=(nh)^{1/2}f_X^{-1}(x)\big\{(\widehat{\mu}_{n,h}-\mu_{n,h})(x)\\&\quad-m(x)(\widehat{f}_{X,n,h}-f_{X,n,h})(x)\big\}+r_{n,h}\end{aligned}\tag{3.2}$$

where $r_{n,h}=o_{L_2}(1)$.

(c) For every x *in* $\mathcal{I}_X$ *and for every integer* $p>1$, $\|\widehat{\mu}_{n,h}(x)-\mu(x)\|_p$ *and* $\|\widehat{m}_{n,h}(x)-m(x)\|_p$ *converge to zero, the bias of the estimators* $\widehat{\mu}_{n,h}(x)$ *and* $\widehat{m}_{n,h}(x)$ *is uniformly approximated by*

$$\begin{aligned}b_{\mu,n,h}(x)&=\mu_{n,h}(x)-\mu(x)=h^2b_\mu(x)+o(h^2),\\ b_\mu(x)&=\frac{m_{2K}}{2}\mu^{(2)}(x)=\frac{m_{2K}}{2}\int y\frac{\partial^2 f_{XY}(x,y)}{\partial x^2}\,dy,\end{aligned}\tag{3.3}$$

$$\begin{aligned}b_{m,n,h}(x)&=m_{n,h}(x)-m(x)=h^2b_m(x)+o(h^2),\\ b_m(x)&=f_X^{-1}(x)\{b_\mu(x)-m(x)b_f(x)\}\\&=\frac{m_{2K}}{2}f_X^{-1}(x)\{\mu^{(2)}(x)-m(x)f_X^{(2)}(x)\},\end{aligned}\tag{3.4}$$

the covariance between $\widehat{\mu}_{n,h}(x)$ *and* $\widehat{f}_{X,n,h}(x)$ *is*

$$\begin{aligned}Cov_{\mu,f_X,n,h}(x)&=(nh)^{-1}\{Cov_{\mu,f_X}(x)+o(1)\},\\ Cov_{\mu,f_X}(x)&=\mu(x)\kappa_2=m(x)f_X(x)\kappa_2\end{aligned}\tag{3.5}$$

and their variance

$$\begin{aligned}v_{\mu,n,h}(x)&=(nh)^{-1}\{\sigma_\mu^2(x)+o(1)\},\\ \sigma_\mu^2(x)&=w_2(x)\kappa_2,\end{aligned}\tag{3.6}$$

$$\begin{aligned}v_{m,n,h}(x)&=(nh)^{-1}\{\sigma_m^2(x)+o(1)\},\\ \sigma_m^2(x)&=\{w_2(x)-m^2(x)f(x)\}\kappa_2 f_X^{-2}(x)\\&=\kappa_2 f_X^{-1}(x)\sigma^2(x).\end{aligned}\tag{3.7}$$

Proof. Note that Condition 3.1 implies that the kernel estimator of f_X is bounded away from zero on $\mathcal{I}_X$ which may be a sub-interval of the support of the variable X. Proposition 2.2 and the almost sure convergence to zero of $\sup_{x\in\mathcal{I}_{X,h}}|\widehat{\mu}_{n,h}-\mu_{n,h}|$, proved by the same arguments as for the density, imply the assertion (a). The bias and the variance are similar for

$\widehat{\mu}_{n,h}(x)$ and $\widehat{f}_{X,n,h}(x)$. For $\widehat{\mu}_{n,h}(x)$, they are a consequence of (b). The first approximation of (b) comes from the expansion

$$\begin{aligned}
\widehat{m}_{n,h}(x) &= \frac{\mu_{n,h}(x)}{f_{X,n,h}(x)} + \frac{(\widehat{\mu}_{X,n,h} - \mu_{X,n,h})(x)}{f_{X,n,h}(x)} \\
&\quad - \frac{\widehat{m}_{n,h}(x)\{\widehat{f}_{X,n,h}(x) - f_{X,n,h}(x)\}}{f_{X,n,h}(x)} \\
&= \frac{\mu_{n,h}(x)}{f_{X,n,h}(x)} + \frac{(\widehat{\mu}_{X,n,h} - \mu_{X,n,h})(x)}{f_{X,n,h}(x)} - \frac{\widehat{\mu}_{n,h}(x)(\widehat{f}_{X,n,h} - f_{X,n,h})(x)}{f^2_{X,n,h}(x)} \\
&\quad + \frac{\widehat{m}_{n,h}(x)(\widehat{f}_{X,n,h} - f_{X,n,h})^2(x)}{f^2_{X,n,h}(x)}, \\
&\quad - \frac{(\widehat{\mu}_{n,h} - \mu_{n,h})(x)(\widehat{f}_{X,n,h} - f_{X,n,h})(x)}{f^2_{X,n,h}(x)},
\end{aligned}$$

the expectation of this equality yields

$$\begin{aligned}
m_{n,h}(x) &= \frac{\mu_{n,h}(x)}{f_{X,n,h}(x)} - E\frac{(\widehat{\mu}_{n,h} - \mu_{n,h})(x)(\widehat{f}_{X,n,h} - f_{X,n,h})(x)}{f^2_{X,n,h}(x)} \\
&\quad + E\frac{\widehat{m}_{n,h}(x)\{\widehat{f}_{X,n,h}(x) - f_{X,n,h}(x)\}^2}{f^2_{X,n,h}(x)} \\
&= \frac{\mu_{n,h}(x)}{f_{X,n,h}(x)} + O((nh)^{-1}) = \frac{\mu_{n,h}(x)}{f_{X,n,h}(x)} + o(h^2) \qquad (3.8)
\end{aligned}$$

uniformly on $\mathcal{I}_X$, for any bounded regression function m. The bias of $\widehat{m}_{n,h}(x)$ is

$$b_{m,n,h}(x) = \Big\{\frac{\mu_{n,h}(x)}{f_{X,n,h}(x)} - m(x)\Big\} + \Big\{m_{n,h}(x) - \frac{\mu_{n,h}(x)}{f_{X,n,h}(x)}\Big\},$$

where the second difference is a $o(h^2)$, using (3.8). A second order Taylor expansion of $f^{-1}_{X,n,h}(x)$ as n tends to infinity leads to

$$\frac{\mu_{n,h}(x)}{f_{X,n,h}(x)} = m(x) + \{b_{\mu,n,h}(x) - m(x)b_{f_X,n,h}(x)\}f_X^{-1}(x) + o(h^2)$$

and the bias of $\widehat{m}_{n,h}(x)$ follows immediatly.

The variance $v_{m,n,h}(x)$ of $\widehat{m}_{n,h}(x)$ is

$$v_{m,n,h}(x) = E\Big\{\widehat{m}_{n,h}(x) - \frac{\mu_{n,h}(x)}{f_{X,n,h}(x)}\Big\}^2 - \Big\{m_{n,h}(x) - \frac{\mu_{n,h}(x)}{f_{X,n,h}(x)}\Big\}^2,$$

where the non random term is a $o(h^4)$, by (3.8). The first term develops using twice the equality $y^{-1} = x^{-1} - (y-x)(xy)^{-1}$

$$\begin{aligned}
f_{X,n,h}(x)\{\widehat{m}_{n,h}(x) - \frac{\mu_{n,h}(x)}{f_{X,n,h}(x)}\} &= \widehat{\mu}_{n,h}(x) - \mu_{n,h}(x) \\
&- m_{n,h}(x)\{\widehat{f}_{X,n,h}(x) - f_{X,n,h}(x)\} \qquad (3.9)\\
&- \frac{\{\widehat{\mu}_{n,h}(x) - \mu_{n,h}(x)\}\{\widehat{f}_{X,n,h}(x) - f_{X,n,h}(x)\}}{f_{X,n,h}(x)} \\
&+ \frac{\widehat{m}_{n,h}(x)\{\widehat{f}_{X,n,h}(x) - f_{X,n,h}(x)\}^2}{f_{X,n,h}(x)},
\end{aligned}$$

so that

$$\begin{aligned}
f^2_{X,n,h}(x)E\{\widehat{m}_{n,h}(x) - \frac{\mu_{n,h}(x)}{f_{X,n,h}(x)}\}^2 &= Var\{\widehat{\mu}_{n,h}(x)\} \\
&+ m^2_{n,h}(x)Var\{\widehat{f}_{X,n,h}(x)\} - 2m_{n,h}(x)Cov\{\widehat{\mu}_{n,h}(x), \widehat{f}_{X,n,h}(x)\} \\
&+ \frac{\pi_{0,2,2}(x)}{f^2_{X,n,h}(x)} + 2\frac{m_{n,h}(x)}{f_{X,n,h}(x)}\pi_{0,1,2}(x) - 2\frac{\pi_{0,2,1}(x)}{f_{X,n,h}(x)} \\
&+ \frac{\pi_{2,0,4}(x)}{f^2_{X,n,h}(x)} + 2\frac{\pi_{1,1,2}(x)}{f_{X,n,h}(x)} - 2m_{n,h}(x)\frac{\pi_{1,0,3}(x)}{f_{X,n,h}(x)} \\
&- 2\frac{\pi_{1,1,3}(x)}{f^2_{X,n,h}(x)},
\end{aligned}$$

where

$$\pi_{k,k',k''}(x) = E\big[\widehat{m}^k_{n,h}(x)\{\widehat{\mu}_{n,h}(x) - \mu_{n,h}(x)\}^{k'}\{\widehat{f}_{X,n,h}(x) - f_{X,n,h}(x)\}^{k''}\big]$$

for $k \geq 0$, $k' \geq 0$ and $k'' \geq 0$. Since $\widehat{m}_{n,h}(x)$ is bounded, Cauchy-Schwarz inequalities and the order of the moments of $\widehat{\mu}_{n,h}(x)$ and $\widehat{f}_{n,h}(x)$ imply that all terms $\pi_{k,k',k''}(x)$ in the above expression are $O((nh)^{-(k'+k'')/2}$ so they are $o((nh)^{-1})$ except the covariance term $\pi_{0,1,1}(x)$. Using the first order expansions of the means $f_{X,n,h}(x) = f_X(x) + O(h^2)$ the mean develops as $m_{n,h}(x) = m(x) + O(h^2)$. It follows that

$$\begin{aligned}
v_{m,n,h}(x) = \{f_X(x)\}^{-2}\big[&Var\{\widehat{\mu}_{n,h}(x)\} + m^2(x)\ Var\{\widehat{f}_{X,n,h}(x)\} \\
&- 2m(x)\ Cov\{\widehat{\mu}_{n,h}(x), \widehat{f}_{X,n,h}(x)\}\big] + o(n^{-1}h^{-1})
\end{aligned}$$

and the convergence to zero of the last term $r_{n,h}$ in (3.2) is satisfied. The other results are obtained by simple calculus. □

The minimax property of the estimator $\widehat{m}_{n,h}$ is established by the same method as for density estimation.

For $p \geq 2$, let

$$w_p(x) = E(Y^p 1_{\{X=x\}}) \tag{3.10}$$

be the p-th moment of Y conditionally on $X = x$. The L_p risk is calculated from the approximation (3.2) of Proposition 3.1 and the next lemmas.

Lemma 3.1. *For $p \geq 2$*

$$\|\widehat{\mu}_{n,h}(x) - \mu_{n,h}(x)\|_p = O((nh)^{-1/p})$$

and

$$\|\widehat{f}_{n,h}^{-1}(x) - f_{X,n,h}^{-1}(x)\|_p = O((nh)^{-1/p}),$$

where the approximations are uniform.

Proof. By the expansion (3.9), Proposition 2.2 extends to $\widehat{\mu}_{n,h}(x) - \mu_{n,h}(x)$ and the moments of order $p \geq 2$ of $\widehat{\mu}_{n,h}(x) - \mu_{n,h}(x)$ and $\widehat{f}_{X,n,h}(x) - f_{X,n,h}$ are $0((nh)^{-1/p})$ which is decreasing as p increases. Let $a_n = f_{n,h}^{-1}\{\widehat{f}_{n,h} - f_{n,h}\}$, then

$$\{\widehat{f}_{n,h}^{-1} - f_{n,h}^{-1}\}^p = f_{n,h}^{-p}\{(1+a_n)^{-1} - 1\}^p = f_{n,h}^{-p}\{\sum_{k\geq 1}(-a_n)^k\}^p$$

and the decreasing order of the moments of the kernel estimator of the density implies

$$E|\sum_{k\geq 1}(-a_n)^k|^p = E|a_n|^p + o(E|a_n|^p).$$

□

The convergence rate of the bandwidth determines the behaviour of the bias term of the process $(nh)^{1/2}(\widehat{m}_{n,h} - m)$, with the following technical results. They generalise Proposition 3.1 to p and $s \geq 2$.

Lemma 3.2. *Under Conditions 2.1, the bias of $\widehat{\mu}_{n,h}(x)$ and $\widehat{m}_{n,h}(x)$ are uniformly approximated as*

$$\begin{aligned} b_{\mu,n,h}(x;s) &= \frac{h^s}{s!} m_{sK} \int y \frac{\partial^s f_{X,Y}(x,y)}{\partial x^s}\, dy + o(h^s), \\ b_{m,n,h}(x;s) &= \frac{h^s}{s!} m_{sK} f_X^{-1}(x)\{\mu^{(s)}(x) - m(x) f_X^{(s)}(x)\} + o(h^s), \end{aligned} \tag{3.11}$$

for $s \geq 2$, and their variances develop as in Proposition 3.1.

Proposition 3.2. *Under Conditions 2.1 and 3.1 with $s = 2$, for every x in $\mathcal{I}_{Xh}$*

$$(nh)^{1/2}(\widehat{m}_{n,h} - m) = (nh)^{1/2} f_X^{-1}\{(\widehat{\mu}_{n,h} - \mu_{n,h}) - m(\widehat{f}_{X,n,h} - f_{X,n,h})\} + (nh^5)^{1/2} b_m + \widehat{r}_{n,h}, \quad (3.12)$$

and the remainder term of (3.12) satisfies

$$\sup_{x \in \mathcal{I}_{X,h}} \|\widehat{r}_{n,h}\|_2 = O((nh)^{-1/2}).$$

Proof. Expanding (3.2) yields

$$\begin{aligned}\widehat{r}_{n,h} &= (nh)^{1/2}(\widehat{f}_{X,n,h}^{-1} - f_X^{-1})\{(\widehat{\mu}_{n,h} - \mu_{n,h}) - m(\widehat{f}_{X,n,h} - f_{X,n,h})\} \\ &+ (nh)^{1/2} \widehat{f}_{X,n,h}^{-1} f_{X,n,h} (f_{X,n,h}^{-1} \mu_{n,h} - m) - (nh^5)^{1/2} b_\mu \\ &= (nh)^{1/2}(\widehat{f}_{X,n,h}^{-1} - f_X^{-1})\{(\widehat{\mu}_{n,h} - \mu_{n,h}) - m(\widehat{f}_{X,n,h} - f_{X,n,h})\} \\ &+ (nh)^{1/2}(\mu_{n,h} f_{X,n,h}^{-1} - m - h^2 b_\mu) \qquad (3.13) \\ &+ (nh)^{1/2}(f_{X,n,h}^{-1} \mu_{n,h} - m) \sum_{k \geq 1} \Big(-\frac{\widehat{f}_{X,n,h} - f_{X,n,h}}{f_{X,n,h}}\Big)^k.\end{aligned}$$

By Lemma 3.1 and Proposition 3.1, the first term is a $O((nh)^{-1/2})$. The second order uniform approximation

$$f_{X,n,h}^{-1}(x)\mu_{n,h}(x) - m(x) = h^2 b_\mu(x) + O(h^4), \quad (3.14)$$

implies that the second term in the sum is a $O(h^4) = O((nh)^{-1})$, as a consequence of Condition 2.1. By Lemma 3.1 and (3.14), the third term is a $O((nh)^{1/2} h^2 (nh)^{-1/2})$, it is therefore a $O((nh)^{-1/2})$. □

For a regression function of class C_s, $s \geq 2$, the L_2-norm of the remainder term $\widehat{r}_{n,h}$ is given by the next proposition.

Proposition 3.3. *Under Conditions 2.1, 2.2 and 3.1, for every $s \geq 2$ the remainder term of (3.12) satisfies the uniform bounds*

$$\sup_{\mathcal{I}_{X,h}} \|\widehat{r}_{n,h}\|_2 = O((nh)^{-1/2}).$$

Proof. For functions f_X and μ in C_s, the risk of $\widehat{r}_{n,h}$ is modified by the bias terms of the previous expansion (3.13). The second term in the approximation (3.14) is replaced by $f_{X,n,h}^{-1}(x)\mu_{n,h}(x) - m(x) = h^s b_\mu(x) + O(h^{s+1})$ and Conditions 2.2 and 3.2, which implies $h^{2s} = O((nh)^{-1})$. By Lemma 3.2, $\sup_x E|\widehat{r}_{n,h}(x)|^2$ is bounded by

$$O(nh)[\{O(h^{2s} + (nh)^{-1})\} O((nh)^{-1}) + O(h^{2(2+s)}) + O(h^{2s}(nh)^{-1})]$$

which is a $O(h^{2s}) + O(h^4) + O(h^{2s}) = O(h^4)$. □

Propositions 2.2, 3.1, Equation (3.2), Propositions 3.2 and 3.3 determine an upper bound for the norm $\|\widehat{m}_{n,h} - m_{n,h}\|_p$ of the estimator of m

$$\begin{aligned}\|\widehat{m}_{n,h} - m_{n,h}\|_p &= \|f_X^{-1}\{(\widehat{\mu}_{n,h} - \mu_{n,h}) - m(\widehat{f}_{X,n,h} - f_{X,n,h})\}\|_p \\ &\quad + O((nh)^{-1/2}\|\widehat{r}_{n,h}\|_p), \\ &\leq 2^{p-1}[\sup_{\mathcal{I}_X} f_X^{-1} \{\|\widehat{\mu}_{n,h} - \mu_{n,h}\|_p \\ &\quad + \sup_{\mathcal{I}_X} |m| \, \|\widehat{f}_{X,n,h} - f_{X,n,h}\|_p\}] + O((nh)^{-1/2}\|\widehat{r}_{n,h}\|_p),\end{aligned}$$

it is therefore a $O((nh)^{-1/2})$. The expression of the L_p-norm $\|\widehat{m}_{n,h} - m_{n,h}\|_p$ is obtained by similar expansions and approximations as in the proof of Proposition 3.1.

Under Conditions 2.1, 2.2, 3.1, for a function μ in C_s and a density f_X in C_r, the bias of $\widehat{m}_{n,h}$ is

$$b_{m,n,h}(x) = f_X^{-1}(x)\{\frac{h^s}{s!}b_\mu(x) - m(x)\frac{h^r}{r!}b_f(x)\} + o(h^{s\wedge r})$$

and its variance does not depend on r and s.

The derivability conditions f_X and $\mu \in C_s$ of 3.1 can be replaced by the condition: f_X and μ belong to a Hölder class $\mathcal{H}_{\alpha,M}$.

Proposition 3.4. *Assume f_X and μ are bounded and belong to $\mathcal{H}_{\alpha,M}$ then the bias of $\widehat{m}_{n,h}(x)$ is bounded by $Mm_{[\alpha]K}h^\alpha/([\alpha]!)f_X^{-1}(x)\{1 + |m(x)|\} + o(h^\alpha)$, by equation (3.2). The optimal bandwidth is $O(n^{1/(2\alpha+1)})$ and the MISE at the optimal bandwidth is $O(n^{\alpha/(2\alpha+1)})$.*

3.3 Optimal bandwidths

The asymptotic mean squared error of $\widehat{m}_{n,h}(x)$, for $p = 2$, is

$$\begin{aligned}(nh)^{-1}\sigma_m^2(x) + h^4 b_m^2(x) &= (nh)^{-1}\kappa_2 f_X^{-2}(x)\{w_2(x) - m^2(x)f(x)\} \\ &\quad + \frac{h^4 m_{2K}^2}{4} f_X^{-2}(x)\{\mu^{(2)}(x) - m(x)f_X^{(2)}(x)\}^2\end{aligned}$$

and its minimum is reached at the optimal bandwidth

$$h_{AMSE}(x) = \{\frac{\kappa_2}{m_{2K}^2}\frac{n^{-1}\{w_2(x) - m^2(x)f(x)\}}{\{\mu^{(2)}(x) - m(x)f_X^{(2)}(x)\}^2}\}^{1/5}$$

where $AMSE(x) = O(n^{-4/5})$. The global mean squared error criterion is the integrated error and it is approximated by

$$\begin{aligned}AMISE &= (nh)^{-1}\kappa_2 \int f_X^{-2}(x)\{w_2(x) - m^2(x)f(x)\}\,dx \\ &\quad + \frac{h^4 m_{2K}^2}{4}\int f_X^{-2}(x)\{\mu^{(2)}(x) - m(x)f_X^{(2)}(x)\}^2\,dx\end{aligned}$$

and the optimal global bandwidth is

$$h_{n,AMISE} = \Big\{\frac{\kappa_2}{m_{2K}^2}\frac{n^{-1}\int f_X^{-1}(x)Var\{Y|X=x\}f(x)\}\,dx}{\int f_X^{-2}(x)\{\mu^{(2)}(x)-m(x)f_X^{(2)}(x)\}^2\,dx}\Big\}^{1/5}.$$

For every $s \geq 2$, the asymptotic quadratic risk of the estimator for a regression curve of class C_s is

$$\begin{aligned} AMSE(x) &= (nh)^{-1}\sigma_m^2(x) + h^{s2}b_{m,s}^2(x) \\ &= (nh)^{-1}\kappa_2 f_X^{-2}(x)\{w_2(x) - m^2(x)f(x)\} \\ &\quad + \frac{h^{2s}}{(s!)^2}m_{sK}^2 f_X^{-2}(x)\{\mu^{(s)}(x) - m(x)f_X^{(s)}(x)\}^2, \end{aligned}$$

its minimum is reached at the optimal bandwidth

$$h_{AMSE}(x) = \Big\{\frac{(s!)^2\kappa_2}{2sm_{sK}^2}\frac{n^{-1}\{w_2(x)-m^2(x)f(x)\}}{\{\mu^{(s)}(x)-m(x)f_X^{(s)}(x)\}^2}\Big\}^{1/(2s+1)}$$

where $AMSE(x) = O(n^{-2s/(2s+1)})$. The global mean squared error criterion is the integrated error and it is approximated by

$$\begin{aligned} AMISE(h,s) &= (nh)^{-1}\kappa_2 \int f_X^{-1}(x)Var\{Y \mid X=x\}\,dx \\ &\quad + \frac{h^{2s}m_{sK}^2}{(s!)^2}\int f_X^{-2}(x)\{\mu^{(s)}(x)-m(x)f_X^{(s)}(x)\}^2\,dx \end{aligned}$$

and the optimal global bandwidth is

$$h_{n,AMISE}(s) = \Big\{\frac{(s!)^2\kappa_2}{2sm_{sK}^2}\frac{n^{-1}\int f_X^{-1}(x)Var\{Y\mid X=x\}\,dx}{\int f_X^{-2}(x)\{\mu^{(s)}(x)-m(x)f_X^{(s)}(x)\}^2\,dx}\Big\}^{1/(2s+1)},$$

and again $AMISE(h_n(s),s) = O(n^{-2s/(2s+1)})$.

In order to estimate the constants of the optimal bandwidths, a nonparametric estimator of the conditional variance of Y are defined as

$$\widehat{V}ar_{n,h}(Y|X=x) = \frac{\sum_{i=1}^n Y_i^2 K_h(x-X_i)}{\sum_{i=1}^n K_{h_2}(x-X_i)} - \widehat{m}_{n,h}^2(x).$$

More generally, the conditional moment of order p, $m_p(x) = E(Y^p|X=x)$ is estimated by

$$\widehat{m}_{p,n,h}(x) = \frac{\sum_{i=1}^n Y_i^p K_h(x-X_i)}{\sum_{i=1}^n K_h(x-X_i)}$$

with a bandwidth $h = h_n$ such that h_n tends to zero and nh_n^2 tends to infinity as n tends to infinity. For every $p \geq 2$, the estimator $\widehat{m}_{p,n,h}$ is also written $\widehat{f}_{n,h}^{-1}\widehat{\mu}_{p,n,h}$, it is a.s. uniformly consistent and approximations similar to those of Propositions 3.1 and 3.2 for the regression curve hold for $m_p(x)$

$$(nh)^{1/2}(\widehat{m}_{p,n,h} - m_{p,n,h}) = (nh^5)^{1/2} b_{p,m} + (nh)^{1/2} f_X^{-1}\{(\widehat{\mu}_{p,n,h} - \mu_{p,n,h}) - m_p(\widehat{f}_{X,n,h} - f_{X,n,h})\} + \widehat{r}_{p,n,h},$$

$$\sup_{x \in \mathcal{I}_{X,h}} \|\widehat{r}_{p,n,h}\|_2 = O((nh)^{-1/2})$$

and for its bias

$$b_{\mu_p,n,h} = \frac{m_{2K} h^2}{2} \int y^p \frac{\partial^2 f_{XY}(\cdot, y)}{\partial x^2}\, dy + o(h^2),$$

$$b_{m_p,n,h} = m_{p,n,h} - m_p = \frac{m_{2K} h^2}{2} f_X^{-1}\{b_{\mu_p,n,h} - m_p b_f\} + o(h^2).$$

The covariance between $\widehat{\mu}_{p,n,h}(x)$ and $\widehat{f}_{X,n,h}(x)$ is $(nh)^{-1} m_p(x) f_X(x) \kappa_2$ and the variances of the estimators of $\mu_p(x)$ and $m_p(x)$ are

$$v_{\mu_p,n,h}(x) = (nh)^{-1}\{w_p(x)\kappa_2 + o(1)\},$$

$$v_{m_p,n,h}(x) = (nh)^{-1}\{\kappa_2 f_X^{-2}(x)\{\sigma^2_{\mu_p}(x) - m_p^2(x) f(x)\} + o(1)\}.$$

The estimators of the derivatives of the regression function m are

$$\begin{aligned}
\widehat{m}^{(1)}_{n,h}(x) &= \frac{\sum_{i=1}^n Y_i K_h^{(1)}(x - X_i)}{\sum_{i=1}^n K_h(x - X_i)} \\
&\quad - \frac{\{\sum_{i=1}^n Y_i K_h(x - X_i)\}\{\sum_{i=1}^n K_h^{(1)}(x - X_i)\}}{\{\sum_{i=1}^n K_h(x - X_i)\}^2}, \\
&= \widehat{f}^{-1}_{X,n,h}(x)\{\widehat{\mu}^{(1)}_{n,h}(x) - \widehat{m}_{n,h}(x) \widehat{f}^{(1)}_{X,n,h}(x)\} \qquad (3.15)
\end{aligned}$$

and all consecutive derivatives of this expression. The first derivatives

$$\widehat{\mu}^{(1)}_{n,h}(x) = n^{-1} \sum_{i=1}^n Y_i K_h^{(1)}(x - X_i)$$

and $\widehat{f}^{(1)}_{X,n,h}(x) = n^{-1} \sum_{i=1}^n K_h^{(1)}(x - X_i)$ converge uniformly on $\mathcal{I}_{X,h}$ to their expectations $\mu^{(1)}_{n,h}(x) = h^{-1} E Y K_h^{(1)}(x - X)$ and $f^{(1)}_{X,n,h}(x)$, respectively, where

$$f^{(1)}_{X,n,h}(x) = f_X^{(1)}(x) + \frac{h^2}{2} m_{2K} f_X^{(3)}(x) + o(h^2),$$

$$\begin{aligned}
\mu^{(1)}_{n,h}(x) &= h^{-1} \int y K_h'(u - x) f_{X,Y}(u, y)\, du\, dy \\
&= (m f_X)^{(1)}(x) + \frac{h^2}{2} m_{2K} (m f_X)^{(3)}(x) + o(h^2),
\end{aligned}$$

then $\widehat{m}_{n,h}^{(1)}$ converges uniformly to $f_X^{-1}(x)\{(mf_X)^{(1)} - mf_X^{(1)}\} = m^{(1)}$, as h tends to zero. The bias of $\widehat{m}_{n,h}^{(1)}(x)$ is

$$\frac{h^2}{2} m_{2K} f_X^{-1}(x)\{(mf_X)^{(3)} - mf_X^{(3)}\}(x).$$

Its variance is obtained by an application of Proposition 3.1 to equation (3.15), its convergence rate is $(nh^3)^{-1}$ (see Appendix A) and the optimal global bandwidth for estimating $m^{(1)}$ follows. For the second derivative

$$\begin{aligned}\widehat{m}_{n,h}^{(2)}(x) = {} & \frac{\sum_{i=1}^n Y_i K_h^{(2)}(x - X_i)}{\sum_{i=1}^n K_h(x - X_i)} \\ & - 2\frac{\{\sum_{i=1}^n Y_i K_h^{(1)}(x - X_i)\}\{\sum_{i=1}^n K_h^{(1)}(x - X_i)\}}{\{\sum_{i=1}^n K_h(x - X_i)\}^2} \\ & - \frac{\{\sum_{i=1}^n Y_i K_h(x - X_i)\}\{\sum_{i=1}^n K_h^{(2)}(x - X_i)\}}{\{\sum_{i=1}^n K_h(x - X_i)\}^2} \\ & + 2\frac{\{\sum_{i=1}^n Y_i K_h(x - X_i)\}\{\sum_{i=1}^n K_h^{(1)2}(x - X_i)\}}{\{\sum_{i=1}^n K_h(x - X_i)\}^3},\end{aligned}$$

the estimators $\widehat{f}_{n,h}^{(2)}$ and $\widehat{\mu}_{n,h}^{(2)}(x) = n^{-1}\sum_{i=1}^n Y_i K_h^{(2)}(x - X_i)$ converge uniformly to $f^{(2)}$ and $\mu^{(2)}$, respectively, with respective biases $\frac{h^2}{2} m_{2K} f_X^{(4)}(x) + o(h^2)$ and $\frac{h^2}{2} m_{2K} \mu^{(4)}(x) + o(h^2)$. The result extends to a general order of derivative $k \geq 1$.

Proposition 3.5. *Under Conditions 2.2 and 3.1 with $nh^{2k+2s+1} = O(1)$, for $k \geq 1$, and functions m and f_X in class $C_s(\mathcal{I}_X)$, the estimator $\widehat{m}_{n,h}^{(k)}$ is an uniformly consistent estimator of the k-order derivative of the regression function, its bias is a $O(h^s)$, and its variance a $O((nh^{2k+1})^{-1})$, the optimal bandwidth is a $O(n^{-1/(2k+2s+1)})$.*

The nonparametric estimator (3.1) is often used in nonparametric time series models with correlated errors. The bias is unchanged and the variance of the estimator depends on the covariances between the observation errors $E(\varepsilon_i\varepsilon_{i+a}) = \beta_a$, for a weakly stationary process $(Y_i)_i$ corresponding to correlated measurements of $Y = m(X) + \varepsilon$. Now the variance σ_f^2 is replaced by $S = \sigma_f^2 + 2\sum_{i\geq 1}\beta_a$ assumed to be finite (Billingsley, 1968). A consistent estimator of S was defined by $\widehat{S}_m = \sum_{i=-m}^m \widehat{\beta}_i$ where the correlation is estimated by the mean correlation error with a mean over the lag between the terms of the product and a sum over observations and $n^{-1}m^2$ tends to zero (Herrmann, Gasser and Kneip, 1992).

3.4 Weak convergence of the estimator

The weak convergence of the process $U_{n,h} = (nh)^{1/2}\{\widehat{m}_{n,h} - m\}\mathcal{I}_{\{\mathcal{I}_{X,h}\}}$ relies on bounds for the moments of its increments which are first proved, as in Lemma 2.2 for the increments of the centered process defined by the kernel estimator, with a kernel having the compact support $[-1, 1]$. For a function or a process φ defined on $\mathcal{I}_{X,h}$, let $\Delta\varphi(x, y) = \varphi(x) - \varphi(y)$.

Lemma 3.3. *Under Conditions 3.1, there exist positive constants C_1 and C_2 such that for every x and y in $\mathcal{I}_{X,h}$ and satisfying $|x - y| \le 2h$*

$$E|\Delta(\widehat{\mu}_{n,h} - \mu_{n,h})(x, y)|^2 \le C_1(nh^3)^{-1}|x - y|^2,$$
$$E|\Delta(\widehat{m}_{n,h} - m_{n,h})(x, y)|^2 \le C_2(nh^3)^{-1}|x - y|^2,$$

if $|x - y| > 2h$, they are $O((nh)^{-1})$ and the estimators at x and y are independent.

Proof. Let x and y in $\mathcal{I}_{X,h}$ such that $|x-y| \le 2h$, $E|\widehat{\mu}_{n,h}(x) - \widehat{\mu}_{n,h}(y)|^2$ develops as the sum $n^{-1}\int w_2(u)\{K_{h_n}(x-u) - K_{h_n}(y-u)\}^2 f(u)\,du + (1 - n^{-1})\{\mu_{n,h_n}(x) - \mu_{n,h_n}(y)\}^2$. For an approximation of the integral, the Mean Value Theorem implies $K_{h_n}(x-u) - K_{h_n}(y-u) = (x-y)\varphi_n^{(1)}(z-u)$ where z is between x and y, and

$$\begin{aligned}\int \{K_{h_n}(x-u) - K_{h_n}(y-u)\}^2 w_2(u) f(u)\,du \\ &= (x-y)^2 \int \varphi_n^{(1)2}(z-u) w_2(u) f(u)\,du \\ &= (x-y)^2 h_n^{-3}\{w_2(x) f(x) \int K^{(1)2} + o(h_n)\}.\end{aligned}$$

Let $|x| \le h_n$ and $|y| \le h_n$, the order of the second moment $E|\widehat{f}_{n,h}(x) - \widehat{f}_{n,h}(y)|^2$ is a $O((x-y)^2(nh_n^3)^{-1})$ if $|x-y| \le 2h_n$ and it is the sum $E\widehat{\mu}^2_{n,h}(x)$ and $\widehat{\mu}^2_{n,h}(y)$ otherwise. This bound and Lemma 2.2 imply the same orders for the estimator of the regression function m. □

Theorem 3.1. *For $h > 0$, the process $U_{n,h} = (nh)^{1/2}\{\widehat{m}_{n,h} - m\}\mathcal{I}_{\{\mathcal{I}_{X,h}\}}$ converges in distribution to $\sigma_m W_1 + \gamma^{1/2} b_m$ where W_1 is a centered Gaussian process on $\mathcal{I}_X$ with variance 1 and covariances zero.*

Proof. For any $x \in \mathcal{I}_{X,h}$ and from the approximation (3.2) of Proposition 3.1 and the weak convergences for $\widehat{\mu}_{n,h} - \mu_{n,h}$ and $\widehat{f}_{X,n,h} - f_{X,n,h}$, the variable $U_{n,h}(x)$ develops as $(nh)^{1/2}\{\widehat{m}_{n,h}(x) - m_{n,h}(x)\} + (nh^5)^{1/2} b_m(x) + o((nh^5)^{1/2})$, and it converges to a non centered distribution $\{W + \gamma^{1/2} b_m\}(x)$ where $W(x)$ is the Gaussian variable with mean zero and variance $\sigma_m^2(x)$. In the same way, the finite dimensional distributions of the process $U_{n,h}$ converge weakly to those of $\{W + \gamma^{1/2} b_m\}$, where W is a Gaussian process with the same distribution as $W(x)$ at x. The covariance matrix $\{\sigma^2(x_k, x_l)\}_{k,l=1,\dots,m}$ between components $W(x_k)$ and $W(x_l)$ of the limiting process is the limit of

$$Cov\{U_{n,h}(x_k), U_{n,h}(x_l)\} = \frac{nh}{f_X(x_k) f_X(x_l)} \big[Cov\{\widehat{\mu}_{n,h}(x_k), \widehat{\mu}_{n,h}(x_l)\}$$
$$- m(x_k) Cov\{\widehat{f}_{X,n,h}(x_k), \widehat{\mu}_{n,h}(x_l)\} - m(x_l) Cov\{\widehat{\mu}_{n,h}(x_k), \widehat{f}_{X,n,h}(x_l)\}$$
$$+ m(x_k) m(x_l) Cov\{\widehat{f}_{X,n,h}(x_k), \widehat{f}_{X,n,h}(x_l)\} + o(1)\big],$$

where the $o(1)$ is deduced from Propositions 3.1, 3.2 and 3.3. For every integers k and l, let $\alpha_h = |x_l - x_k|/(2h)$ and $v = \{(x_l + x_k)/2 - s\}/h$ be in $[0, 1]$, hence $h^{-1}(x_k - s) = v - \alpha$ and $h^{-1}(x_l - s) = v + \alpha$. By a Taylor expansion in a neighborhood of $(x_l + x_k)/2$, the integral of the first covariance term develops as

$$Cov\{\widehat{\mu}_{n,h}(x_k), \widehat{\mu}_{n,h}(x_l)\}$$
$$= n^{-1} h^{-1} w_2\Big(\frac{x_k + x_l}{2}\Big) f_X\Big(\frac{x_k + x_l}{2}\Big) \int K(v - \alpha_h) K(v + \alpha_h)\, dv$$
$$+ o(n^{-1} h^{-1})$$

and zero otherwise, with the notation (3.10). Similar expansions are satisfied for the other terms of the covariance. Using the following approximations for $|x_k - x_l| \le 2h$: $w_2(\{x_k + x_l\}/2) = w_2(x_k) + o(1) = w_2(x_l) + o(1)$ and $f_X(\{x_k + x_l\}/2) = f_X(x_k) + o(1) = f_X(x_l) + o(1)$, the covariance of $U_{n,h}(x_k)$ and $U_{n,h}(x_l)$ is approximated by

$$\frac{Var(Y|X = x_k) + Var(Y|X = x_l)}{f_X(x_k) + f_X(x_l)} \mathcal{I}_{\{0 \le \alpha_h < 1\}} \int K((v - \alpha_h) K(v + \alpha_h)\, dv.$$

Due to the compactness of the support of K, the covariance is zero if $\alpha_h \ge 1$. For $x_k \ne x_l$, α_h tends to infinity as h tends to zero and $I\{0 \le \alpha_h < 1\}$ tends to zero as n tends to infinity, therefore the covariance of $U_{n,h}(x_k)$ and $U_{n,h}(x_l)$ is equal to $\delta_{k,l} + o(1)$, where $Var\{U_{n,h}(x_k)\}$ is defined in Proposition 3.1.

The tightness of the sequence $\{U_{n,h}\}$ on $\mathcal{I}_{X,h}$ will follow from (i) the tightness of $\{U_{n,h}(a)\}$ and (ii) a bound of the increments $E\big|U_{n,h}(x_2) - U_{n,h}(x_1)\big|^2$ for $|x_2 - x_1| < 2h$. For condition (i), let $\eta > 0$ and $c > \gamma^{1/2}|b_m(a)| + \big(2\eta^{-1}\sigma^2(a)\big)^{1/2}$, then

$$\Pr\{|U_{n,h}(a)| > c\} \le \Pr\{(nh)^{1/2}|(\widehat{m}_{n,h} - m_{n,h})(a)| + (nh)^{1/2}|b_{n,h}(a)| > c\}$$
$$\le \frac{Var\{(nh)^{1/2}(\widehat{m}_{n,h} - m_{n,h})(a)\}}{\{c - (nh)^{1/2}|b_{n,h}(a)|\}^2}$$

and for n sufficiently large

$$\Pr\{|U_{n,h}(a)| > c\} \le \frac{\sigma^2(a)}{\{c - \gamma^{1/2}|b_m(a)|\}^2} + o(1) < \eta.$$

The process $U_{n,h}$ is written $W_{n,h} + (nh)^{1/2} b_{n,h}$ where

$$(b_{n,h}(x) - b_{n,h}(y))^2 \le kh^{2s}(x-y)^{2s} = O((nh)^{-1})(x_1 - x_2)^{2s}$$

and $W_{n,h} = (nh)^{1/2}(\widehat{m}_{n,h} - m_{n,h})$. From Lemma 3.3, there exists a constant C_W such that $|x-y| \le 2h$ entails $E\{W_{n,h}(x) - W_{n,h}(y)\}^2 \le C_W h^{-2}|x-y|^2$, which implies the tightness of the process $U_{n,h}$ and its weak convergence to a continuous Gaussian process defined on $\mathcal{I}_X$. □

Note that the tightness of the process implies the existence of a constant $c_\eta > 0$ such that

$$\Pr\{\sup_{\mathcal{I}_{X,h}} |\sigma_m^{-1}(U_{n,h} - \gamma^{1/2} b_m) - W_1| > c_\eta\} \xrightarrow{P} 0.$$

The limiting distribution of the process $U_{n,h}$ does not depend on the bandwidth h, so one can state the following corollary.

Corollary 3.1. $\sup_{h>0: nh^{2s+1} \to \gamma} \sup_{\mathcal{I}_{X,h}} \sigma_m^{-1}|U_{n,h} - \gamma^{1/2} b_m|$ *converges in distribution to* $\sup_{\mathcal{I}_X} |W_1|$.

An uniform confidence interval for the regression curve m is deduced as for the density.

Let X be a variable defined in a subset $\mathcal{I}_X$ of $\mathbb{R}^d$, the regression function m is estimated using a multivariate kernel K defined on $[-1,1]^d$ and $K_h(x) = h^{-d}K(h^{-d}x)$, for $x = (x_1, \ldots, x_d)$ in $\mathcal{I}_{X,h}$. The bias is unchanged and the rates of the moments $p \ge 2$ are modified by the dimension d

$$Var\{\widehat{m}_{n,h}(x)\} = (nh^d)^{-1}\kappa_2 Var\{Y|X=x\} f_X^{-1}(x) + o((nh^d)^{-1}),$$

and $\|\widehat{f}_{n,h}(x) - f_{n,h}(x)\|_p = 0((nh^d)^{-1/p})$. The local and global errors $MISE_n(h)$ are $O(h^{2s}) + O((nh^d)^{-1})$, they are minimal at the optimal bandwidths of order $O(n^{-1/(2s+d)})$ where the MISE reaches the minimal order $O(n^{-2s/(2s+d)})$. The weak convergence of Theorem 3.1 and its corollary still hold with the rate $(nh^d)^{1/2}$.

3.5 Estimation of a regression curve by local polynomials

The regression function m is approximated by a Taylor expansion of order k, for every s in a neighborhood $\mathcal{V}_{x,h}$ of a fixed x, with radius h,

$$m(s) = m(x) + (s-x)m'(x) + \ldots + \frac{(s-x)^p}{p!} m^{(p)} + o((s-x)^p). \quad (3.16)$$

This expansion is a local polynomial regression where the derivatives at x are considered as parameters. Estimating the derivatives by the derivatives of the estimator $\widehat{m}_{n,h}$ yields an estimator having a variance sum of terms of different orders, its main term is the variance of $\widehat{m}_{n,h}$.

Let $(H_{k,h})_k$ be a square integrable orthonormal basis of real functions with respect to the distribution function of X, with support $\mathcal{V}_{x,h}$ for h converging to zero. Let $\delta_{k,l}$ be the Dirac indicator $\delta_{k,l}$ of equality for k and l, $k, l \geq 0$. Equation (3.16) is also written

$$m(s) = \sum_{k=0}^{p} \theta_k(x) H_k(s-x) + o((s-x)^p) = m_p(x) + o((s-x)^p)$$

for s in $\mathcal{V}_{x,h}$, and the properties of the functional basis entail

$$E\{H_k(X-x)H_l(X-x)\} = \int H_k(s-x)H_l(s-x)\,dF(s) = \delta_{k,l},\ k,l \geq 0.$$

In the regression model $E(Y|X) = m(X)$

$$\begin{aligned} \theta_k(x) &= E\{YH_k(X-x)\} = E\{H_k(X-x)m(X)\},\ k \geq 1 \\ m(x) &= E\{YH_0(X-x)\} = E\{H_0(X-x)m(X)\}. \end{aligned}$$

For fixed x, $\theta_k(x)$ is considered as a constant parameter. This expansion is an extension of the kernel smoothing if the functional basis has regularity properties. The nonparametric regression function is approximated by an expansion on the first p elements of the basis and its projections satisfy $\theta_k(x) = \int m(s)H_k(x-s)\,dF(s)$. The estimation of the parameters is performed by the projection of the observations of Y onto the first p elements of the orthonormal basis. Let $(X_i, Y_i)_{i=1,\ldots,n}$ be a sample for the regression variables (X, Y), so that $Y_i = m(X_i) + \varepsilon_i$ where ε_i is an observation error having a finite variance $\sigma^2 = E\{Y - m(X)\}$ and such that $E(\varepsilon|X) = 0$. An estimator of the parameter is defined as the empirical conditional mean of the projection of Y onto the space generated by the basis. For $k \geq 1$,

$$\widehat{\theta}_{k,n}(x) = n^{-1} \sum_{i=1}^{n} Y_i H_k(X_i - x)$$

is therefore a consistent estimator of θ_k. Its conditional variance is $n^{-1}\{E(Y^2|X)H_k^2(X-x) - \theta_k^2\}\{1 + o(1)\}$.

This approach may be compared to the local polynomials defined by minimizing the local smoothed empirical mean squared error

$$ASE(x) = \sum_{i=1}^{n} \{Y_i - m_p(X_i, \theta)\}^2 K_h(X_i - x).$$

This provides an estimator of θ with components satisfying

$$\sum_{i=1}^{n}\{Y_i - m_p(X_i,\theta)\}H_k(X_i - x)K_h(X_i - x) = 0.$$

They are solution of a system of linear equations and θ_{nk} is approximated by

$$\frac{\sum_{i=1}^{n} Y_i H_k(X_i - x)K_h(X_i - x)}{\sum_{i=1}^{n} K_h(X_i - x)}$$

if the orthogonality of the basis entails that $EH_k(X-x)H_l(X-x)K_h(X-x)$ convergences to zero as h tends to zero, for every $k \neq l \leq p$. This estimator is consistent and its behaviour is further studied by the same method as the estimator of the nonparametric regression.

A multidimensional regression function $m(X_1,\ldots,X_d)$ can be expanded in sums of univariate regression functions $E(Y \mid X_k = x)$ and their interactions like a nonparametric analysis of variance if the regression variables $(X_1,\ldots,X_d)$ generate orthogonal spaces generated. The orthogonality is a necessary condition for the estimation of the components of this expansion since

$$\begin{aligned} E\{YK_h(x_k - X_k)\} &= \int E(Y \mid X = x)\, F_X(dx_1,\ldots,x_{k-1}, \\ &\qquad\qquad x_{k+1},\ldots,x_d) + o(1) \\ &= m(x_k)f_k(x_k) + o(1), \\ E\{YK_h(x_k - X_k)K_h(x_l - X_l)\} &= C_K m(x_k,x_l) f_{X_k,X_l}(x_k,x_l) + o(1), \end{aligned}$$

where $C_K = \int K(u)K(v)\,du\,dv$, and $E\{YK_h(x_k - X_k)K_h(x_l - X_l)\}$ $f^{-1}_{X_k,X_l}(x_k,x_l)$ can be factorized or expanded as a sum of regression functions only if X_k and X_l belong to orthogonal spaces. The orthogonalisation of the space generated by a vector variable X can be performed by a preliminary principal component analysis providing orthogonal linear combinations of the initial variables.

3.6 Estimation in regression models with functional variance

Consider the nonparametric regression model with an observation error funtion of the regression variable X, $Y = m(X) + \sigma(X)\varepsilon$ defined by (1.7), with $E(\varepsilon|X) = 0$ and $Var(\varepsilon|X) = 1$. The variance $\sigma(x)^2 = E[\{(Y - m(X)\}^2|X = x]$ is assumed to be continuous and it is estimated by a localisation of the empirical error in a neighborhood of x

$$\widetilde{\sigma}^2_{n,h,\delta}(x) = \frac{\sum_{i=1}^n \{Y_i - \widehat{m}_{n,h}(X_i)\}^2 1\{X_i \in V_\delta(x)\}}{\sum_{i=1}^n 1\{X_i \in V_\delta(x)\}}$$

or by smoothing it with a kernel density

$$\widehat{\sigma}^2_{n,h,\delta}(x) = \frac{\sum_{i=1}^n \{Y_i - \widehat{m}_{n,h}(X_i)\}^2 K_\delta(x - X_i)}{\sum_{i=1}^n K_\delta(x - X_i)}. \tag{3.17}$$

The estimator is denoted $\widehat{\sigma}^2_{n,h,\delta}(x) = \widehat{f}^{-1}_{X,n,\delta}(x)\widehat{S}_{n,h,\delta}(x)$, with

$$\begin{aligned}\widehat{S}_{n,h,\delta}(x) &= n^{-1}\sum_{i=1}^n \{Y_i - \widehat{m}_{n,h}(X_i)\}^2 K_\delta(x - X_i) \\ &= \int \{y - \widehat{m}_{n,h}(s)\}^2 K_\delta(x - s)\, d\widehat{F}_{X,Y,n}(s, y).\end{aligned}$$

The mean of $\widehat{S}_{n,h,\delta}(x)$ is denoted $S_{n,h,\delta}(x)$. By the uniform consistency of $\widehat{m}_{n,h}$, $\widehat{S}_{n,h,\delta}$ converges uniformly to S as n tends to infinity, with h and δ tending to zero. At X_j, it is written $\widehat{S}_{n,h,\delta}(X_j) = n^{-1}\sum_{i\neq j}\{Y_i - \widehat{m}_{n,h}(X_i)\}^2_{n,h} K_\delta(X_j - X_i) + o((nh)^{-1})$. The rate of convergence of δ_n to zero is governed by the degree of derivability of the variance function σ^2.

Condition 3.2. For a density f_X in $C_r(\mathcal{I}_X)$ and a function μ in $C_s(\mathcal{I}_X)$ and a variance σ^2 in $C_k(\mathcal{I}_X)$, with $k, s, r \geq 2$, the bandwidth sequences $(\delta_n)_n$ and $(h_n)_n$ satisfy

$$\delta_n = O(n^{-1/(2k+1)}), \quad h_n = O(n^{-1/\{2(s\wedge r)+1\}}),$$

as n tends to infinity.

Proposition 3.6. *Under Conditions 2.1, 2.2 and 3.1, for every function μ in C_s, density f_X in C_r and variance function σ^2 in C_k,*

$$E\{Y - \widehat{m}_{nh}(x)\}^2 = \sigma^2(x) + O(h^{2(s\wedge r)}) + O((nh)^{-1}),$$

the bias of the estimator $\widehat{S}_{n,h,\delta}(x)$ of $\sigma^2(x)$ defined by (3.17) is

$$\begin{aligned}\beta_{n,h,\delta}(x) = b^2_{m,n,h}(x) f_X(x) + \sigma^2_{m,n,h}(x) f_X(x) + \frac{\delta^{2k}}{(k!)^2}(\sigma^2(x) f_X(x))^{(2)} \\ + o(\delta^{2k} + h^{2(s\wedge r)} + (nh)^{-1})\end{aligned}$$

and its variance is written $(n\delta)^{-1}\{v_{\sigma^2} + o(1)\}$ with $v_{\sigma^2}(x) = \kappa_2 Var\{(Y - m(x))^2 | X = x\}$. The process $(n\delta)^{1/2}(\widehat{\sigma}^2_{n,h,\delta} - \sigma^2 - \beta_{n,h,\delta})$ converges weakly to a Gaussian process with mean zero, variance v_{σ^2} and covariances zero.

Proof. Using Proposition 2.2 and Lemma 3.2, the mean squared error for $\widehat{m}_{nh}$ at x is $E[\{Y - \widehat{m}_{nh}(x)\}^2 \mid X = x]$ and it is expanded as $\sigma^2(x) + b^2_{m,n,h}(x) + \sigma^2_{m,n,h}(x) + E[\{Y - m(x)\}\{m(x) - \widehat{m}_{nh}(x)\} \mid X = x]$ where the last term is zero. For the variance of $\widehat{S}_{n,h,\delta}(x)$, the fourth conditional moment $E[\{Y - \widehat{m}_{nh}(x)\}^4(x) \mid X = x]$ is the conditional expectation of $\{(Y - m(x)) + (m - m_{nh})(x) + (m_{nh} - \widehat{m}_{nh})(x)\}^4$ and it is expanded in a sum of $\sigma_4(x) = E\{Y - m(x))^4 \mid X = x\}$, a bias term $b^4_{m,n,h}(x) = O(h^{8(s\wedge r)})$, $E(m_{nh} - \widehat{m}_{nh})(x)\}^4 = O((nh)^{-1})$ by Proposition 3.1, and products of squared terms the main of which being $\sigma^2(x)\|\widehat{m}_{nh} - m\|_2^2(x)$ of order $O((nh)^{-1}) + O(h^{4(s\wedge r)})$, and the others being smaller. The variance of $\widehat{S}_{n,h,\delta}(x)$ follows.

Moreover, for every $i \neq j \leq n$ and for every function ψ in C_2 and integrable with respect to F_X, $E\psi(X_j)K^2_\delta(X_i - X_j) = \int \psi(x)K^2_\delta(x - x')\,dF_X(x)\,dF_X(x')$ equals $\kappa_2 E\psi(X) + o(\delta^2)$ and the main term of the variance does not depend on the bandwidth δ. □

The bandwidths h_n and δ_n appear in the bias and the variance, therefore the mean squared error for the variance is minimum under Condition 3.2.

Note that the function m which achieves the minimum of the empirical mean squared error for the model $V_{n,h}(x) = n^{-1}\sum_{i=1}^n K_h(x - X_i)\{Y_i - m(x)\}^2$ is the estimator $\widehat{m}_{n,h}$ (3.1) and $V_{n,h}(x)$ converges in probability $\sigma(x)$. In a parametric regression model with a Gaussian error having a constant variance, $V_n(x) = n^{-1}\sum_{i=1}^n\{Y_i - m(x)\}^2$ is the sufficient statistic for the estimation of the parameters of m. In a Gaussian regression model with a functional variance $\sigma^2(x)$, each term of the sum defining the error is normalized by a different variance $\sigma(X_i)$ and the sufficient statistic for the estimation of parameters of the function m is the weighted mean square error

$$V_{w,n}(\theta) = n^{-1}\sum_{i=1}^n \sigma^{-1}(X_i)\{Y_i - m_\theta(X_i)\}^2.$$

For a nonparametric regression function, an empirical local mean weighted squared error is defined as

$$V_{w,n,h}(x) = n^{-1}\sum_{i=1}^n w(X_i)\{Y_i - m(x)\}^2 K_h(x - X_i)$$

with $w(x) = \sigma^{-1}(x)$. A weighted estimator of the nonparametric regression curve m is then defined as

$$\widehat{m}_{w,n,h}(x) = \frac{\sum_{i=1}^n w(X_i)Y_i K_h(x - X_i)}{\sum_{i=1}^n w(X_i)K_h(x - X_i)}, \tag{3.18}$$

if the variance is known, it achieves the minimum of $V_{w,n,h}(x)$. With an unknown variance, minimizing the weighted squared error leads to the estimator built with its estimator $\widehat{w}_n = \widehat{\sigma}^{-1}_{n,h_n,\delta_n}$, using (3.17)

$$\widehat{m}_{\widehat{w}_n,n,h}(x) = \frac{\sum_{i=1}^n \widehat{w}_n(X_i)Y_iK_h(x-X_i)}{\sum_{i=1}^n \widehat{w}_n(X_i)K_h(x-X_i)}. \tag{3.19}$$

The uniform consistency of $\widehat{w}_{n,h}$ implies $\sup_{\mathcal{I}_{n,h}} |\widehat{m}_{\widehat{w}_n,n,h} - m_w|$ tends to zero as n tends to infinity.

Assuming that σ belongs to $C_2(\mathcal{I}_X)$, the convergence results for $\widehat{m}_{n,h}$ in Propositions 3.1 or (3.2) adapt to the estimator (3.18), with $\mu_w = w\mu$ instead of μ and $w(x)f_{X,Y}(x,y)$ instead of $f_{X,Y}(x,y)$. The approximation (3.2) is unchanged, hence the bias and the variance of the weighted estimator $\widehat{m}_{w,n,h}$ are

$$b_{m,w,n,h}(x) = \frac{h^s m_{sK}}{s!w(x)f_X(x)}\{(mwf_X)^{(s)}(x) - m(x)(wf_X)^{(s)}(x)\} + o(h^s),$$

$$v_{m,w,n,h}(x) = v_{m,n,h}(x).$$

In the approximations of Propositions 3.2 and 3.3, the order of convergence of $\sup_{x\in\mathcal{I}_{X,h}} \|\widehat{r}_{n,h}\|_2$ is not modified and the weak convergence of Theorem 3.1 is fulfilled for the process $(nh)^{1/2}\{\widehat{m}_{w,n,h} - m\}\mathcal{I}_{\{\mathcal{I}_{X,h}\}}$, with the modified bias and variance.

With an estimated weight, the mean of the numerator $\widehat{\mu}_{\widehat{w}_n,n,h}(x)$ is $E\widehat{w}_n(X)m(X)K_h(x-X)$ and it equals $\int E\widehat{w}_n(y)m(y)K_h(x-y)f_X(y)\,dy$ since $\widehat{\sigma}^2_{n,h_n,\delta_n}(X_i)$ is equivalent to the estimator of the variance (at X_i) calculated from the observations without X_i. With an empirical weight $\widehat{w}_n(x) = \psi(\widehat{\sigma}^2_{n,h_n,\delta_n}(x))$, the mean of the numerator of the estimator (3.19) is then $EN_n(x) = Ew(X)m(X)K_h(x-X) + E\{(\widehat{\sigma}^2_{n,h_n,\delta_n} - \sigma^2)(X)\psi'(\sigma^2(X))m(X)K_h(x-X)\}\{1+o(1)\}$ and the bias of the numerator of (3.19) is modified by adding $m(x)f_X(x)\beta_{n,h,\delta}(x)\psi'(\sigma^2(x))$ to the bias of the expression with a fixed weight w. In the same way, the expectation of the denominator is $ED_n(x) = w(X)K_h(x-X) + E\{(\widehat{\sigma}^2_{n,h_n,\delta_n} - \sigma^2)(X)\psi'(\sigma^2)(X)K_h(x-X)\}\{1+o(1)\}$ and it is approximated by $f_X(x)\{w(x) + \beta_{n,h,\delta}(x)\psi'(\sigma^2(x))$. Using the approximation (3.2) of Proposition 3.1, the first order approximation of the bias of (3.19) is identical to $b_{m,w,n,h}(x)$. The variances of each term are

$$VarN_n(x) = \frac{\kappa_2}{nh}E\{\widehat{w}_n^2(x)E(Y^2 \mid X=x)f_X(x)\} + o((nh)^{-1}),$$

$$VarD_n(x) = \frac{\kappa_2}{nh}E\{\widehat{w}_n^2(x)f_X(x)\} + o((nh)^{-1}),$$

$$Var\widehat{m}_{\widehat{w}_n,n,h}(x) = \frac{\kappa_2}{nhw^2(x)f_X(x)}Var(\widehat{w}_n(x)Y \mid X=x) + o((nh)^{-1}).$$

The variance of the estimator with an empirical weight is therefore modified by a random factor in the variance of Y and a normalization by $w(x)$. The convergence rates are not modified.

3.7 Estimation of the mode of a regression function

The mode of a real regression function m on $\mathcal{I}_X$ is

$$M_m = \sup_{\mathcal{I}_X} m(x). \tag{3.20}$$

The mode M_m of a regular regression function is estimated by the mode of a regular estimator of the function, $\widehat{M}_{m,n,h} = M_{\widehat{m}_{n,h}}$. Under Conditions 2.1-3.1, the regression function is locally concave in a neighborhood $\mathcal{N}_M$ of the mode and its estimator has the same property for n sufficiently large, by the uniform consistency of $\widehat{m}_{n,h}$, hence $m^{(1)}(M_m) = 0$, $m^{(2)}(M_m) < 0$, $\widehat{m}^{(1)}_{n,h}(\widehat{M}_{m,n,h}) = 0$ and $\widehat{M}_{m,n,h}$ converges to M_m in probability. A Taylor expansion of $m^{(1)}$ at the estimated mode implies

$$(\widehat{M}_{m,n,h} - M_m) = \{m^{(2)}(M_m)\}^{-1}\{m^{(1)}(\widehat{M}_{m,n,h}) - \widehat{m}^{(1)}_{n,h}(\widehat{M}_{m,n,h})\} + o(1).$$

The weak convergences of the process $(nh^3)^{1/2}(\widehat{m}^{(1)}_{n,h} - m^{(1)})$ (Proposition 3.5) determines the convergence rate of $(\widehat{M}_{m,n,h} - M_m)$ as $(nh^3)^{-1/2}$ and it implies the asymptotic behaviour of the estimator $\widehat{M}_{m,n,h}$.

Proposition 3.7. *Under Conditions 2.1, 2.2 and 3.1, $(nh^3)^{1/2}(\widehat{M}_{m,n,h} - M_m)$ converges weakly to a centered Gaussian variable with finite variance $m^{(2)-2}(M_m)Var\widehat{m}^{(1)}_{n,h}(M_m)$.*

If the regression function belongs to $C_3(\mathcal{I}_X)$, the bias of $m^{(1)}(\widehat{M}_{m,n,h})$ is deduced from the bias of the process $\widehat{m}^{(1)}_{n,h}$ defined by (3.15), it equals

$$Em^{(1)}(\widehat{M}_{m,n,h}) = -\frac{h^2}{2} m_{2K} f_X^{-1}(x)\{(mf_X)^{(3)} - mf_X^{(3)}\}(M_m) + o(h^2)$$

and does not depend on the degree of derivability of the regression function m. All results are extended for the search of the local maxima and minima of the function m which are local maxima of $-m$. The maximization of the function on the interval $\mathcal{I}_X$ is then replaced by sequential maximizations or minimizations.

3.8 Estimation of a regression function under censoring

Consider the nonparametric regression (1.6) where the variable Y is right-censored by a variable C independent of (X, Y) and the observed variables are (X, Y^*, δ) where $Y^* = Y \wedge C$ and $\delta = 1_{\{Y \le C\}}$. Let $F_{Y|X}$ denote the distribution function of Y conditionally on X. The regression function $m(x) = E(Y \mid X = x) = \int y F_{Y|X}(dy; x)$ is estimated using an estimator of the conditional density of Y given X under right-censoring. Extending the results of Section 2.8 to the nonparametric regression, the conditional distribution function $F_{Y|X}$ defines a cumulative conditional hazard function

$$\Lambda_{Y|X}(y; x) = \int 1_{\{s \le y\}} \{1 - F_{Y|X}(s; x)\}^{-1} F_{Y|X}(ds; x),$$

conversely the function $\Lambda_{Y|X}$ uniquely defines the conditional distribution function as

$$1 - F_{Y|X}(y; x) = \exp\{-\Lambda^c_{Y|X}(y; x)\} \prod_{z>y} \{1 - \Delta\Lambda_{Y|X}(z^-; x)\},$$

where $\Lambda^c_{Y|X}$ is the continuous part of $\Lambda_{Y|X}$ and $\prod_s \{1 - \Delta\Lambda(s)\}$ its right-continuous discrete part. Let

$$N_n(y; x) = \sum_{1 \le i \le n} K_h(x - X_i)\delta_i 1_{\{Y_i \le y\}},\; Y_n(y; x) = \sum_{1 \le i \le n} K_h(x - X_i) 1_{\{Y_i^* \ge y\}}$$

be the counting processes related to the observations of the censored variable Y^*, with regressors in a neighborhood $\mathcal{V}_h(x)$ of x, and let $J_n(y; x)$ be the indicator of $Yn(y; x) > 0$. The process $M_n(y; x) = N_n(y; x) - \int_{-\infty}^{y} Y_n(s; x)\, d\Lambda_{Y|X}(s; x)$ is a centered martingale with respect to the filtration generated by the observed processes up to y^-, conditionally on regressors in $\mathcal{V}_h(x)$. The functions $\Lambda_{Y|X}$ and $F_{Y|X}$ are estimated by

$$\widehat{\Lambda}_{Y|X,n,h}(y; x) = \int 1_{\{s \le y\}} \frac{J_n(s; x) N_n(ds; x)}{Y_n(s; x)},$$

$$\widehat{F}_{Y|X,n,h}(y; x) = 1 - \prod_{Y_i \le y} \{1 - \Delta\widehat{\Lambda}_{Y|X,n,h}(Y_i; x)\},$$

the estimator $\widehat{\Lambda}_{Y|X,n,h}$ is unbiased and $\widehat{F}_{Y|X,n,h}$ is the Kaplan-Meier estimator for distribution function of Y conditional on $\{X = x\}$. The regression function m is then estimated by

$$\begin{aligned}\widehat{m}_{n,h}(x) &= \int y \widehat{F}_{Y|X,n,h}(dy; x) \\ &= \sum_{i=1}^{n} Y_i \{1 - \widehat{F}_{Y|X,n,h}(Y_i^-; x)\} \frac{J_n(Y_i; x)}{Y_n(Y_i; x)}.\end{aligned}$$

The estimators satisfy $\sup_{\mathcal{I}_X\times\mathcal{I}}|\widehat{\Lambda}_{Y|X,n,h}-\Lambda_{Y|X}|$, $\sup_{\mathcal{I}_{X,Y}}|\widehat{F}_{Y|X,n,h}-F_{Y|X}|$ and $\sup_{\mathcal{I}_X}|\widehat{m}_{n,h}-m|$ converge in probability to zero as n tends to infinity, for every compact subinverval $\mathcal{I}$ of $\mathcal{I}_Y$. For every $y \le \max Y_i^*$, the conditional Kaplan-Meier estimator, given x in $\mathcal{I}_{X,n,h}$, still satisfies

$$\frac{F_{Y|X}-\widehat{F}_{Y|X,n,h}}{1-F_{Y|X}}(y;x)=\int_{-\infty}^{y}\frac{1-\widehat{F}_{Y|X,n,h}(s^-;x)}{1-F_{Y|X}(s;x)}\,d(\widehat{\Lambda}_{Y|X,n,h}-\Lambda_{Y|X})(s;x). \tag{3.21}$$

The mean of this integral with respect to a centered martingale is zero so the conditional Kaplan-Meier estimator and $\widehat{\Lambda}_{Y|X,n}$ are unbiased estimators. The bias of the estimator of the regression function for censored variables Y is then a $O(h^2)$.

3.9 Proportional odds model

Consider a regression model with a discrete response variable Y corresponding to a categorization of an unobserved continuous real variable Z in a partition $(I_k)_{k\le K}$ of its range, with the probabilities $\Pr(Z\in I_k)=\Pr(Y=k)$. With a regression variable X and intervals $I_k=(a_{k-1},a_k)$, the cumulated conditional probabilities are

$$\pi_k(X)=\Pr(Y\le k\mid X)=\Pr(Z\le a_k\mid X),$$

and $E\pi_K(X)=1$. The proportional odds model is defined through the logistic model for the probabilities $\pi_k(X)=p(a_k-m(X))$, with the logistic probability $p(y)=\exp(y)/\{1-\exp(y)\}$ and a regression function m. This model is equivalent to $\pi_k(X)\{1-\pi_k(X)\}^{-1}=\exp\{a_k-m(X)\}$ for every function π_k such that $0<\pi_k(x)<1$ for every x in $\mathcal{I}_X$ and for $1\le k<K$. This implies that the odds-ratio for the observations (X_i,Y_i) and (X_j,Y_j) with Y_i and Y_j in the same class does not depend on the class

$$\frac{\pi_k(X_i)\{1-\pi_k(X_j)\}}{\{1-\pi_k(X_i)\}\pi_k(X_j)}=\exp\{m(X_j)-m(X_i)\},$$

for every $k=1,\dots,K$, this is the proportional odds model.

For $k=1,\dots K$, let $p_k(x)=(\pi_k-\pi_{k-1})(x)=\Pr(Y=k\mid X=x)$. Assuming that $p_1(x)>0$ for every x in $\mathcal{I}_X$, the conditional distribution of the discrete variable is also determined by the conditional probabilities $\alpha_k(x)=P(Y=k|X=x)P^{-1}(Y=1|X=x)$. Equivalently

$$P(Y=k|X=x)=\frac{\alpha_k(x)}{1+\sum_{j=1}^{K}\alpha_j(x)},\ k=1,\dots,K,$$

with the constraint $\sum_{k=1}^{K} P(Y = k|X = x) = 1$ for every x. This reparametrization of the conditional probabilities α_k is not restrictive, though it is called the logistic model.

Estimating first the support of the regression variable reduces the number of unknown parameters to $2(K-1)$, the thresholds of the classes and their probabilities, for $k \leq K-1$, in addition to the nonparametric regression function m. The probability functions $\pi_k(x)$ are estimated by the proportions $\widehat{\pi}_{n,k}(x)$ of observations of the variable Y in class k, conditionally on the regressor value x. Let

$$U_{ik} = \log \frac{\widehat{\pi}_{n,k}(X_i)}{1-\widehat{\pi}_{n,k}(X_i)}, \; i = 1, \ldots, n,$$

calculated from the observations $(X_i, Y_i)_{i=1,\ldots,n}$ such that $Y_i = k$. The variations of the regression function m between two values x and y are estimated by

$$\widehat{m}_{n,h}(x) - \widehat{m}_{n,h}(y) = K^{-1} \sum_{k=1}^{K} \left\{ \frac{\sum_{i=1}^{n} U_{ik} K_h(X_i - x)}{\sum_{i=1}^{n} K_h(X_i - x)} - \frac{\sum_{i=1}^{n} U_{ik} K_h(X_i - y)}{\sum_{i=1}^{n} K_h(X_i - y)} \right\}.$$

This estimator yields an estimator for the derivative of the regression function, $\widehat{m}_{n,h}^{(1)}(x) = \lim_{|x-y|\to 0}(x-y)^{-1}\{\widehat{m}_{n,h}(x) - \widehat{m}_{n,h}(y)\}$ wich is written as the mean over the classes of the derivative estimator (3.15) with response variables U_{ik}. Integrating the mean derivative provides a nonparametric estimator of the regression function m. The bounds of the classes cannot be identified without observations of the underlying continuous variable Z, thus the odds ratio allows to remove the unidentifiable parameters from the model for the observed variables.

With a regression multidimensional variable X, the single-index model or a transformation model (Chapter 7) reduce the dimension of the variable and fasten the convergence of the estimators.

3.10 Estimation for the regression function of processes

Consider a continuously observed stationary and ergodic process $(Z_t)_{t\in[0,T]} = (X_t, Y_t)_{t\in[0,T]}$ with values in $\mathcal{I}_{XY}$, and the regression model $Y_t = m(X_t) + \sigma(X_t)\varepsilon_t$ where $(\varepsilon_t)_{t\in[0,T]}$ is a conditional Brownian motion such that $E(\varepsilon_t \mid X_t) = 0$ and $E(\varepsilon_t\varepsilon_s \mid X_t \wedge X_s) = E\{(\varepsilon_t \wedge \varepsilon_s)^2 \mid X_t \wedge X_s) = 1$. The ergodicity property is expressed by (2.13) or (2.16) for the bivariate process Z. The regression function m is estimated on an interval $\mathcal{I}_{X,Y,T,h}$ by the kernel estimator

$$\widehat{m}_{T,h}(x) = \frac{\int_0^T Y_s K_h(x - X_s)\, ds}{\int_0^T K_h(x - X_s)\, ds}. \tag{3.22}$$

Its numerator is denoted

$$\widehat{\mu}_{T,h}(x) = \frac{1}{T}\int_0^T Y_s K_h(x - X_s)\, ds$$

and its denominator is $\widehat{f}_{X,T,h}(x)$. The mean of $\widehat{\mu}_{T,h}(x)$ and its limit are respectively

$$\mu_{T,h}(x) = \int_{\mathcal{I}_{XY}} y K_h(x - u)\, dF_{XY}(u, y),$$
$$\mu(x) = \int_{\mathcal{I}_{XY}} y f_{XY}(x, y)\, dy = f_X(x) m(x).$$

Under Conditions 2.1-2.2 and 3.1, the bias of $\widehat{\mu}_{T,h}(x)$ is

$$b_{\mu,T,h}(x) = \int_{\mathcal{I}_{XY,T}} y K_h(x-u)\, dF_{XY}(u, y) - \mu(x) = \frac{h_T^s}{s!} m_{sK} \mu^{(s)}(x) + o(h_T^s),$$

its variance is expressed through the integral of the covariance between $Y_s K_h(X_s - x)$ and $Y_t K_h(X_t - x)$. For $X_s = X_t$, the integral on the diagonal $\mathcal{D}_X$ of $\mathcal{I}_{X,T}^2$ is a $(Th_T)^{-1})\kappa_2 w_2(x) + o((Th_T)^{-1})$ and the integral outside the diagonal denoted $I_o(T)$ is expanded using the ergodicity property (2.13). Let $\alpha_h(u, v) = |u - v|/2h_T$

$$\begin{aligned}
I_o(T) &= \int_{[0,T]^2} \int_{\mathcal{I}_{XY}^2 \setminus \mathcal{D}_X} y_1 y_2 K_h(u - x) K_h(v - x) dF_{Z_s, Z_t}(u, y_1, v, y_2)\, \frac{ds}{T}\frac{dt}{T} \\
&= (Th_T)^{-1}\Big\{\int_{\mathcal{I}_X} \int_{\mathcal{I}_{X \setminus \{u\}}} \int_{-1/2}^{1/2} K(z - \alpha_h(u, v)) K(z + \alpha_h(u, v))\, dz \\
&\qquad \mu(u)\mu(v) d\pi_u(v)\, dF_X(u)\Big\}\{1 + o(1)\}.
\end{aligned}$$

For every fixed $u \neq v$, $\alpha_{h_T}(u, v)$ tends to infinity as h_T tends to zero, then the integral $\int_{-1/2}^{1/2} K(z - \alpha_h(u, v)) K(z + \alpha_h(u, v))\, dz$ tends to zero with h_T. If $|u - v| = O(h_T)$, this integral does not tend to zero but the transition probability $\pi_u(v)$ tends to zero as h_T tends to zero, therefore the integral $I_o(T)$ is a $o((Th_T)^{-1})$ as T tends to infinity. The L_p-norm of the estimator satisfies $\|\widehat{\mu}_{T,h}(x) - \mu_{T,h}(x)\|_p = O((Th_T)^{-1/p})$ under the ergodicity condition for k-uplets of the process Z (2.16) and the approximation (3.2) is also satisfied for the estimator $\widehat{m}_{T,h}$. It follows that its bias, for $s \geq 2$, and its variance are approximated by

$$\begin{aligned} b_{m,T,h}(x;s) &= h_T^s b_m(x;s) + o(h_T^s), \\ b_m(x;s) &= f_X^{-1}(x)\{b_\mu(x) - m(x)b_f(x)\} \\ &= \frac{m_{sK}}{s!} f_X^{-1}(x)\{\mu^{(s)}(x) - m(x)f_X^{(s)}(x)\}, \\ v_{m,T,h}(x) &= (Th_T)^{-1}\{\sigma_m^2(x) + o(1)\}, \\ \sigma_m^2(x) &= \kappa_2 f_X^{-1}(x) Var(Y \mid X = x) \end{aligned}$$

and the covariance between $\widehat{m}_{T,h}(x)$ and $\widehat{m}_{T,h}(y)$ tends to zero. The mean squared error of the estimator at x for a marginal density in C_s is then

$$\begin{aligned} MISE_{T,h_T}(x) &= (Th_T)^{-1})\kappa_2 f_X^{-1}(x) Var(Y \mid X = x) \\ &\quad + h_T^{2s} b_m^2(x;s) + o((Th_T)^{-1}) + o(h_T^{2s}) \end{aligned}$$

and the optimal local and global bandwidths minimizing the mean squared (integrated) errors are $O(T^{1/(2s+1)})$

$$h_{AMSE,T}(x) = \left\{\frac{1}{T}\frac{\sigma_m^2(x)}{2sb_m^2(x;s)m(x)}\right\}^{1/(2s+1)}$$

and, for the asymptotic mean integrated squared error criterion

$$h_{AMISE,T} = \left\{\frac{1}{T}\frac{\int \sigma_m^2(x)\,dx}{2s\int b_m^2(x;s)m(x)\,dx}\right\}^{1/(2s+1)}.$$

With the optimal bandwidth rate, the asymptotic mean (integrated) squared errors are $O(T^{2s/(2s+1)})$. The same expansions as for the variance $\widehat{\mu}_{T,h}(x)$ and $\widehat{f}_{X,T,h}(x)$ in Section 2.10 prove that the finite dimension distributions of the process $(Th_T)^{1/2}(\widehat{f}_{T,h} - f - b_{T,h})$ converge to those of a centered Gaussian process with mean zero, covariances zero and variance $\kappa_2 f(x)$ at x. Lemma 3.3 generalizes and the increments $E\{\widehat{f}_{T,h}(x) - \widehat{f}_{T,h}(y)\}^2$ are approximated as $E|\Delta(\widehat{m}_{n,h} - m_{n,h})(x,y)|^2 = O(|x-y|^2(Th_T^3)^{-1})$ for every x and y in $\mathcal{I}_{X,h}$ such that $|x-y| \leq 2h_T$. Then the process $(Th_T)^{1/2}\{\widehat{m}_{T,h} - m\}\mathcal{I}_{\{\mathcal{I}_{X,T}\}}$ converges weakly to $\sigma_m W_1 + \gamma^{1/2} b_m$ where W_1 is a centered Gaussian process on $\mathcal{I}_X$ with variance 1 and covariances zero.

3.11 Exercises

(1) Detail the proof for the approximations of the biases and variances of Proposition 3.1.

(2) Suppose Y is a binary variable with $P(Y|X = x) = p(x)$ and express the bias and the variance of the estimator of the nonparametric probability function p.
(3) Consider a discrete variable with values in an infinite countable set. Define an estimator of the function m under suitable conditions and give the expression of its bias and variance.
(4) Define nonparametric estimators for the bias of the function m and its variance.
(5) Define the optimal bandwidths for the estimation of the function μ and its first order derivative.
(6) Detail the expression of $\|\widehat{m}_{n,h}(x) - m(x)\|_p$ using the orders of the norms established in Section 3.2.
(7) Detail the expressions of the bias and the second order approximation of the variance of $\widehat{\sigma}^2_{n,h,\delta}(x)$ in Proposition 3.6.
(8) Let $F_{Y|X}(y; x) = \Pr(Y \leq y \mid X \leq x)$ be the distribution function of Y conditionally on X and

$$\widehat{F}_{Y|X,n,h}(y; x) = n^{-1} \sum_{i=1}^{n} 1_{\{Y_i \leq y\}} H_h(X_i - x)$$

be a smooth estimator of the conditional distribution function (see Exercise 2.11-(6)). Find the expression of the bias and the variance of $\widehat{F}_{Y|X,n,h}(x)$.

Chapter 4

Limits for the variable bandwidths estimators

4.1 Introduction

The pointwise mean squared error for a density or regression function reaches its minimum at a bandwidth function varying in the domain of the variable X. The question of the behaviour of the estimator of density and regression functions with a varying bandwidth is then settled. All results of Chapters 2 and 3 are modified by this function. Consider a density or a regression function of class $C_s(\mathcal{I}_X)$. Let $(h_n)_n$ be a sequence of functional bandwidths in $C_1(\mathcal{I}_X)$, converging uniformly to zero and uniformly bounded away from zero on $\mathcal{I}_X$. In order to have an optimal bandwidth for the estimation of functions of class C_2, the functional sequence is assumed to satisfy an uniform convergence condition for the uniform norm $\|h_n\|$.

Condition 4.1. There exists a strictly positive function h in $C_1(\mathcal{I}_X)$, such that $\|h\|$ is finite and $\|nh_n^{2s+1} - h\|$ tends to zero as n tends to infinity.

Under this condition, the bandwidth is uniformly approximated as

$$h_n(x) = n^{-1/(2s+1)} h^{1/(2s+1)}(x) + o(n^{-1/(2s+1)}).$$

The increasing intervals $\mathcal{I}_{X,h_n}$ are now defined with respect the uniform norm of the function h_n by $\mathcal{I}_{X,h_n} = \{s \in \mathcal{I}_X; [s - \|h_n\|, s + \|h_n\|] \in \mathcal{I}_X\}$. The main results of the previous chapters are extended to kernel estimators with functional bandwidth sequences satisfying this convergence rate. That is the case of the kernel estimators built with estimated optimal local bandwidths calculated from independent observations.

The second point of this chapter is the definition of an adaptative estimator of the bandwidth, when the degree of derivability of the density varies in its domain of definition, and the behaviour of the estimator of the density with an adaptative estimator. In Chapter 2, the optimal density was obtained under the assumption that the degree of smoothness of the density is known and constant on the interval of the observations. The last assumption flattens the estimated curve by the use of a too large bandwidth in areas with smaller derivability order, the above variable bandwidth $h_n(x)$ does not solves that problem. The cross-validation method allows to define a global bandwidth without knowledge of the class of the density. Other adaptative methods are based the maximal variations of the estimator as the bandwidth varies in a grid $\mathcal{D}_n$ corresponding to a discretization of the possible domain of the bandwidth according to the order of regularity of the density. It can be performed globally or pointwisely.

4.2 Estimation of densities

Let us consider the random process $U_{n,h_n}(x) = (nh_n(x))^{1/2}\{\widehat{f}_{n,h_n(x)}(x) - f(x)\}$ for x in $\mathcal{I}_{X,h_n}$. Under Conditions 2.1 and 4.1, $\sup_{\mathcal{I}} |\widehat{f}_{n,h_n}(x) - f(x)|$ converges *a.s.* to zero for every compact subinterval $\mathcal{I}$ of $\mathcal{I}_{X,h_n}$ and $\|\widehat{f}_{n,h_n}(x) - f(x)\|_p$ tends to zero, as n tends to infinity. The bias of $\widehat{f}_{n,h_n(x)}(x)$ is $b_{n,h_n}(x) = \frac{1}{2}h_n^2(x)m_{2K}f^{(2)}(x) + o(\|h_n\|^2)$, its variance is $Var\{\widehat{f}_{n,h_n(x)}(x)\} = (nh_n(x))^{-1}\kappa_2 f(x) + o((n^{-1}\|h_n^{-1}\|)$ and $\|\widehat{f}_{n,h_n(x)}(x) - f_{n,h_n(x)}(x)\|_p = 0((n^{-1}\|h_n^{-1}\|)^{1/p})$.

Under Conditions 2.1-4.1, for a density of class $C_s(\mathcal{I}_X)$ and for every x in $\mathcal{I}_{X,h}$, the moments of order $p \geq 2$ are unchanged and the bias of $\widehat{f}_{n,h_n(x)}(x)$ is modified as

$$b_{n,h_n}(x;s) = \frac{h_n^s(x)}{s!}m_{sK}f^{(s)}(x) + o(\|h_n\|^s).$$

The MISE and the optimal local bandwidth are similar to those of Chapter 2 using these expressions.

For every u in $[-1,1]$, let α_n and v in $[-1,1]$, $|u|$ in $[0, \{x + h_n(x)\} \wedge \{y + h_n(y)\}]$ be defined by

$$\begin{aligned} \alpha_n(x,y,u) &= \frac{1}{2}\{(u-x)h_n^{-1}(x) - (u-y)h_n^{-1}(y)\}, \qquad (4.1)\\ v = v_n(x,y,u) &= \frac{1}{2}\{(u-x)h_n^{-1}(x) + (u-y)h_n^{-1}(y)\}\\ &= \frac{1}{2h_n(x)h_n(y)}[\{h_n(x)+h_n(y)\}u - xh_n(y) - yh_n(x)], \end{aligned}$$

$$u = u_n(x,y,v) = \{h_n(x)+h_n(y)\}^{-1}\{xh_n(y)+yh_n(x)+2vh_n(x)h_n(y)\},$$
$$z_n(x,y) = \{h_n(x)+h_n(y)\}^{-1}\{xh_n(y)+yh_n(x)\},$$
$$\delta_n(x,y) = 2h_n(x)h_n(y)\{h_n(x)+h_n(y)\}^{-1} = o(1),$$

hence $\alpha_n(x,y,u)$ is also denoted $\alpha_n(x,y,v)$.

Lemma 4.1. *The covariance of $\widehat{f}_{n,h}(x)$ and $\widehat{f}_{n,h}(y)\}$ equals*

$$\frac{2}{n\{h_n(x)+h_n(y)\}}\{f(z_n(x,y))\int K(v-\alpha_n(v))K(v+\alpha_n(v))\,dv$$
$$+\,\delta_n(x,y)f^{(1)}(z_n(x,y))\int vK(v-\alpha_n(v))K(v+\alpha_n(v))\,dv + o(\|h_n\|)\}.$$

Proof. The integral

$$EK_{h_n(x)}(x-X)K_{h_n(y)}(y-X) = \int K_{h_n(x)}(x-u)K_{h_n(y)}(y-u)f_X(u)\,du$$

is expanded changing the variable u in v and it equals

$$\frac{2}{h_n(x)+h_n(y)}\int K(v-\alpha_n(v))K(v+\alpha_n(v))f(u_n(x,y,v))\,dv$$
$$= \frac{2}{h_n(x)+h_n(y)}\{f(z_n(x,y))\int K(v-\alpha_n(v))K(v+\alpha_n(v))\,dv$$
$$+\,\delta_n(x,y)f^{(1)}(z_n(x,y))\int vK(v-\alpha_n(v))K(v+\alpha_n(v))\,dv + o(\|h_n\|)\}. \quad \square$$

Lemma 4.2. *For functions of class $C_s(\mathcal{I}_X)$, $s \geq 1$, and under Conditions 3.1 and 4.1, for every x and y in $\mathcal{I}_{X,h_n}$ the mean variation of $\widehat{f}_{n,h_n}$ between x and y has the order $O(|x-y|)$ and its mean squared variation for $|xh_n^{-1}(x)-yh_n^{-1}(y)| \leq 1$ are $E|\widehat{f}_{n,h_n}(x)-\widehat{f}_{n,h}(y)|^2 = O(n^{-1}\|h_n^{-1}\|^3(x-y)^2)$. Otherwise, it is a $O(n^{-1}\|h_n^{-1}\|)$ and the variables $\widehat{f}_{n,h}(x)$ and $\widehat{f}_{n,h}(y)$ are independent.*

Proof. By the Mean Value Theorem, for every x and y in $\mathcal{I}_{X,h}$ there exists s between x and y such that $|f_{n,h_n}(x) - f_{n,h_n}(y)| = |x-y|f^{(1)}(s)$ and

$$|f_{n,h_n}(x) - f_{n,h_n}(y)| \leq |x-y|\|f^{(1)}\|.$$

Let $z = \lim_n z_n(x,y)$ defined in (4.1). The expectation of $|\widehat{f}_{n,h}(x) - \widehat{f}_{n,h}(y)|^2$ develops as $n^{-1}\int\{K_{h_n(x)}(x-u) - K_{h_n(y)}(y-u)\}^2 f(u)\,du + (1-n^{-1})\{f_{n,h_n}(x) - f_{n,h_n}(y)\}^2$.

Using the notations (4.1), the first term of this sum is expanded as

$$S_{1n} = \frac{1}{nh_n(x)h_n(y)\{h_n(x)+h_n(y)\}} \int \{h_n(x)K(v-\alpha_n(v)) \\ -h_n(y)K(v+\alpha_n(v))\}^2 f(z_n(v))\,dv.$$

The derivability of the bandwidth functions implies

$$\frac{1}{h_n(x)h_n(y)} \int \{h_n(x)K(v-\alpha_n) - h_n(y)K(v+\alpha_n)\}^2 f(z_n)\,dv$$
$$\leq 2[\frac{h_n(x)}{h_n(y)} \int \{K(v-\alpha_n) - K(v+\alpha_n)\}^2 f(z_n)\,dv$$
$$+ \frac{\{h_n(x)-h_n(y)\}^2}{h_n(x)h_n(y)} \int K^2(v-\alpha_n) f(z_n)\,dv],$$
$$S_{1n} \leq \frac{2}{n\{h_n(x)+h_n(y)\}}[\frac{h_n(x)}{h_n(y)} f(z) \int \{K(v-\alpha_n) - K(v+\alpha_n)\}^2\,dv$$
$$+ (x-y)^2 \frac{h_n^{(1)2}(\eta(x-y))}{h_n(x)h_n(y)} \int K^2(v-\alpha_n) f(z_n)\,dv],$$

where η lies in $(-1,1)$, by the Mean Value Theorem, $h_n^{(1)2}(\eta)$ and $h_n(x)h_n(y)$ have the same order, and

$$\int \{K(v-\alpha_n) - K(v+\alpha_n)\}^2\,dv = 4\alpha_n^2 \int K^{(1)2}(v)\,dv = O(|x-y|^2\,\|h_n^{-1}\|^2).$$

It follows that $S_{1n} = O(n^{-1}\|h_n^{-1}\|\,|x-y|^2\,\|h_n^{-1}\|^2)$. Since $h_n^{-1}(x)|x|$ and $h_n^{-1}(y)|y|$ are bounded by 1, the order of $E(n\|h_n^{-1}\|^{-1})^{1/2}|\widehat{f}_{n,h}(x) - \widehat{f}_{n,h}(y)|^2 = O((x-y)^2)$ if $|xh_n(y) - yh_n(x)| \leq h_n(y)h_n(x)$, otherwise $\widehat{f}_{n,h}(x)$ and $\widehat{f}_{n,h}(y)$ are independent and it is a sum of variances. □

Theorem 4.1. *Under the conditions, for a density f of class $C_s(\mathcal{I}_X)$ and a varying bandwidth sequence such that $n\|h_n\|^{2s+1}$ converges to $\|h\|$, the process*

$$U_{n,h_n}(x) = (nh_n(x))^{1/2}\{\widehat{f}_{n,h_n(x)} - f(x)\} I\{x \in \mathcal{I}_{X,\|h_n\|}\}$$

converges weakly to the process defined on $\mathcal{I}_X$ as $W_f(x) + h^{1/2}(x)b_f(x)$, where W_f is a continuous centered Gaussian process with covariance $\sigma_f^2(x)\delta_{\{x,x'\}}$ between $W_f(x)$ and $W_f(x')$.

Proof. The weak convergence of the variable $U_{n,h}(x)$ is a consequence to the L_2-convergence of $(nh_n(x))^{1/2}\{\widehat{f}_{n,h_n}(x) - f(x) - (nh_n^{2s+1}(x))^{1/2}b_f(x)\}$ to $\kappa_2 f(x)$. In the same way, the finite dimensional distributions of the process $U_{n,h}$ converge weakly to those of a centered Gaussian process. The quadratic variations of the bias $\{f_{n,h_n(x)}(x) - f(x) - f_{n,h_n(y)}(y) + f(y)\}^2$ are bounded by

$$|\int K(z)\{f(x + h_n(x)z) - f(x) - f(y + h_n(y)z) - f(y)\}\, dz|^2$$
$$= \frac{m_{sK}}{s!}\|h_n\|^{2s}[\{\frac{h(x)}{\|h\|}\}^{2s} f^{(s)}(x) - \{\frac{h(y)}{\|h\|} f^{(s)}(y)\}^{2s}]^2$$

and it is a $O(\|h_n\|^{2s}|x-y|^2)$. This bound and Lemma 4.2 imply that the mean of the squared variations of the process $U_{n,h}$ on small intervals are $O(|x-y|^2)$, therefore the process $U_{n,h}$ is tight, so it converges weakly to a centered Gaussian process. The covariance of the limiting process at x and y is the limit of the covariance between $U_{n,h}(x)$ and $U_{n,h}(y)$ and it equals $\lim_n nh_n^{1/2}(x)h_n^{1/2}(y)Cov\{\widehat{f}_{n,h}(x), \widehat{f}_{n,h}(y)\}$. The covariance of $\widehat{f}_{n,h}(x)$ and $\widehat{f}_{n,h}(y)$ is approximated by $n^{-1}\int K_{h_n(x)}(x-u)K_{h_n(y)}(y-u)f(u)du$, for $x \neq y$ it develops as

$$\frac{1_{\{0 \le \alpha_n < 1\}}}{n\{h_n(x) + h_n(y)\}} f(z_n(x,y)) \int K(v - \alpha_n)K(v + \alpha_n)dv$$
$$+ o(n^{-1}(h_n(x) + h_n(y))^{-1}).$$

As n tends to infinity, $h_n(x)h_n^{-1}(y)$ and $h_n(y)h_n^{-1}(x)$ are bounded and $1_{\{0 \le \alpha_n < 1\}}$ tends to zero for every $x \neq y$, hence the covariance of $U_{n,h}(x)$ and $U_{n,h}(y)\}$ converges to zero as n tends to infinity. □

The optimal bandwidth for estimating a density has an order between n^{-1} and $n^{-1/(2\alpha+1)}$ with $\alpha > 1/4$, under the conditions nh_n tends to infinity and f belongs locally to $\mathcal{H}_\alpha$ with $\alpha > 1/4$, or C_α with $\alpha \ge 2$. All estimators of the bias of a density depend on its regularity through the constant of the bias and the exponent of h and it cannot be directly estimated without knowledge of α. The bandwidth minimizing the mean squared error of the estimator $\widehat{f}_{n,h}(x)$ is bounded by

$$MSE_{n,h}(x,\alpha) = Var\widehat{f}_{n,h}(x)\{1 + (2\alpha)^{-1}\} \tag{4.2}$$

with an order of smoothness $\alpha > 1/4$, so only the lower bound of the degree is necessary to obtain a bound of the MSE. As the variance of $\widehat{f}_{n,h(x)}(x)$ does not depend on the class of f, it can be estimated using a bandwidth function h_2 such that $n\|h_2\|$ tends to zero and $n\|h_2\|^2$ tends to infinity, by $\widehat{V}ar_{n,h_2}\widehat{f}_{n,h}(x) = (nh(x))^{-1}\kappa_2\widehat{f}_{n,h_2(x)}(x)$. Let $\widehat{MSE}_{n,h,a_n}(x)$ be the estimator of $MSE_{n,h,a_n}(x)$ obtained by plugging the estimator of $Var\widehat{f}_{n,h}(x)$. It can be compared with the bootstrap estimator of the mean squared error $MSE^*_{n,h}(x) = Var^*\widehat{f}_{n,h}(x) + B^{*2}(\widehat{f}_{n,h})(x)$ calculated from a bootstrap sample of independent variables having the distribution $\widehat{F}_n$. This estimator and the bootstrap estimator $Var^* \widehat{f}_{n,h}(x)$ yield an estimator of α, by equation (4.2). An optimal local bandwidth can then be estimated from the estimator of α. The choice of the bandwidth function h_2 relies on the same procedure and the optimal estimator $\widehat{h}_n(x)$ requires iterations of this procedure, starting to an empirical bandwidth calculated from a discretization of its range. Adaptative estimators of the bandwidth were previously defined using empirical thresholds for the variations of the estimator of the density according to the bandwidth, however constants in the thresholds were chosen by numerical recursive procedures.

Another variable bandwidth kernel estimator is defined with a bandwidth function of the variables X_i rather than x

$$\widehat{f}_{X,n,h_n}(x) = \frac{1}{n} \sum_{i=1}^{n} K_{h_n(X_i)}(x - X_i).$$

Its mean is $Ef_{X,n,h_n}(x) = EK_{h_n(X)}(x - X) = \int K_{h_n(y)}(x - y) f_X(y)\, dy$ and its limit is $f_x(x)$, approximating y by x in the integral. Its bias and variance are not expanded as above, the bandwidth at y is now developed as $h_n(y) = h_n(x)\{1 - zh_n^{(1)}(x)\} + o(\|h_n\|^2)$ where $x - y = h_n(y)z$, hence

$$\begin{aligned} f_X(y) &= f_X(x - h_n(x)z + h_n(x)h_n^{(1)}(x)z + o(\|h_n\|^2) \\ &= f_X(x) - h_n(x)zf_X^{(1)}(x) + \frac{1}{2}h_n(x)z^2\{h_n(x)f_X^{(2)}(x) \\ &\quad + 2h_n^{(1)}(x)f_X^{(1)}(x)\} + o(\|h_n\|^2) \end{aligned}$$

and the bias of the estimator is

$$b_{\widehat{f}_{X,n,h}}(x) = \frac{m_{2K}}{2} h_n(x)\{h_n(x)f_X^{(2)}(x) + 2h_n^{(1)}(x)f_X^{(1)}(x)\} + o(\|h_n\|^2).$$

Its variance is

$$Var\widehat{f}_{X,n,h_n}(x) = n^{-1}\{\int K^2_{h_n(y)}(x - y) f_X(y)\, dy - E^2 \widehat{f}_{X,n,h_n}(x)\},$$

$$\begin{aligned} \int K^2_{h_n(y)}(x - y)\, dy &= \int K^2(z) f_X(x - h_n(x)z - h_n(x)h_n^{(1)}(x)z)\, dz \\ &\quad + o(\|h_n\|^2) \\ &= m_{2K}\{f_X(x) - h_n(x)f^{(1)}(x) \int zK^2(z)\, dz + o(\|h_n\|) \end{aligned}$$

the first order approximation of the variances are identical and their second order approximation have the opposite sign.

4.3 Estimation of regression functions

Let us consider the variable bandwidth kernel estimator $\widehat{m}_{n,h_n(x)}(x)$ of the regression function m and the random process related to the estimated regression function $U_{m,n,h_n}(x) = (nh_n(x))^{1/2}\{\widehat{m}_{n,h_n}(x) - m(x)\}\mathcal{I}_{\{x\in\mathcal{I}_{X,\|h_n\|}\}}$. Conditions 2.1 and 4.1 for kernel estimators of densities with variable bandwidth are supposed to be satisfied in addition to Conditions 3.1 for kernel estimators of regression functions. Then $\sup_{x\in\mathcal{I}_{X,\|h_n\|}} |\widehat{m}_{n,h_n(x)}(x) - m(x)|$ converges a.s. to zero with the uniform approximations

$$m_{n,h_n(x)}(x) = \frac{\mu_{n,h_n(x)}(x)}{f_{X,n,h_n(x)}(x)} + O((n\|h_n\|)^{-1}),$$

$$\begin{aligned}(nh_n(x))^{1/2}\{\widehat{m}_{n,h_n(x)} - m_{n,h_n(x)}\}(x) &= \big(nh_n(x)\big)^{1/2}\big\{(\widehat{\mu}_{n,h_n(x)} \\ &- \mu_{n,h_n(x)})(x) - m(x)(\widehat{f}_{X,n,h_n(x)} - f_{X,n,h_n(x)})(x)\big\}f_X^{-1}(x) + r_{n,h_n(x)},\end{aligned}$$

where $r_{n,h_n} = o_{L_2}(1)$, uniformly.
For every x in $\mathcal{I}_{X,\|h_n\|}$ and for every integer $p > 1$, $\|\widehat{m}_{n,h_n(x)}(x) - m(x)\|_p$ converges to zero, the bias of the estimator $\widehat{m}_{n,h_n(x)}(x)$ is uniformly approximated by

$$\begin{aligned}b_{m,n,h_n(x)}(x) &= m_{n,h_n(x)}(x) - m(x) = h_n(x)^2 b_m(x) + o(\|h_n\|^2),\\ b_m(x) &= f_X^{-1}(x)\{b_\mu(x) - m(x)b_f(x)\}\\ &= \frac{m_{2K}}{2} f_X^{-1}(x)\{\mu^{(2)}(x) - m(x)f_X^{(2)}(x)\},\end{aligned}$$

and its variance is deduced from (3.7)

$$\begin{aligned}v_{m,n,h_n(x)}(x) &= (nh_n(x))^{-1}\{\sigma_m^2(x) + o(1)\},\\ \sigma_m^2(x) &= \kappa_2 f_X^{-2}(x)\{w_2(x) - m^2(x)f(x)\}.\end{aligned}$$

For a regression function and a density f_X in class $C_s(\mathcal{I}_X)$, $s \geq 2$, and under Conditons 2.2, the bias of $\widehat{m}_{n,h_n(x)}(x)$ is uniformly approximated by

$$b_{m,n,h_n(x)}(x;s) = \frac{h_n^s(x)}{s!} m_{sK} f_X^{-1}(x)\{\mu^{(s)}(x) - m(x)f_X^{(s)}(x)\} + o(\|h_n\|^s)$$

and its moments are not modified by the degree of derivability. For every x in $\mathcal{I}_{X,\|h_n\|}$

$$\begin{aligned}(nh_n(x))^{1/2}(\widehat{m}_{n,h_n} - m)(x) &= (nh_n(x))^{1/2} f_X^{-1}(x)\{(\widehat{\mu}_{n,h_n(x)} - \mu_{n,h_n(x)})\\ &\quad - m(\widehat{f}_{X,n,h_n(x)} - f_{X,n,h_n(x)})\}(x)\\ &\quad + (nh_n(x)^{2s+1})^{1/2} b_m(x) + \widehat{r}_{n,h_n(x)}(x),\end{aligned}$$

and $\sup_{x\in\mathcal{I}_{X,\|h_n\|}} \|\widehat{r}_{n,h_n(x)}\|_2 = O((n\|h_n\|)^{-1/2})$. The asymptotic mean squared error of $\widehat{m}_{n,h}(x)$ is

$$(nh_n(x))^{-1}\sigma_m^2(x) + h_n(x)^4 b_m^2(x) = (nh_n(x))^{-1}\kappa_2\{w_2(x) - m^2(x)f(x)\} + \frac{h_n^4(x)m_{2K}^2}{4} f_X^{-2}(x)\{\mu^{(2)}(x) - m(x)f_X^{(2)}(x)\}^2 f_X^{-2}(x)$$

and its minimum is reached at the optimal local bandwidth

$$h_{n,AMSE}(x) = \{\frac{\kappa_2}{m_{2K}^2} \frac{n^{-1}\{w_2(x) - m^2(x)f(x)\}}{\{\mu^{(2)}(x) - m(x)f_X^{(2)}(x)\}^2}\}^{1/5}$$

where $AMSE(x) = O(n^{-4/5})$. For every $s \geq 2$, the asymptotic quadratic risk of the estimator for a regression curve of class C_s is

$$\begin{aligned} AMSE(x) &= (nh_n(x))^{-1}\sigma_m^2(x) + h_n^{2s}(x)b_{m,s}^2(x) \\ &= (nh_n(x))^{-1}\kappa_2 f_X^{-2}(x)\{w_2(x) - m^2(x)f(x)\} \\ &\quad + \frac{h_n^{2s}(x)}{(s!)^2} m_{sK}^2 f_X^{-2}(x)\{\mu^{(s)}(x) - m(x)f_X^{(s)}(x)\}^2, \end{aligned}$$

its minimum is reached at the optimal bandwidth

$$h_{n,AMSE}(x) = \{\frac{(s!)^2\kappa_2}{2sm_{sK}^2} \frac{n^{-1}\{w_2(x) - m^2(x)f(x)\}}{\{\mu^{(s)}(x) - m(x)f_X^{(s)}(x)\}^2}\}^{1/(2s+1)}$$

where $AMSE(x) = O(n^{-2s/(2s+1)})$.

The covariance of $\widehat{m}_{n,h_n}(x)$ and $\widehat{m}_{n,h_n}(y)$ is calculated as for Theorems 3.1 and 4.1 and it is a $o(1)$ for every $x \neq y$.

Lemma 4.3. *The covariance of $\widehat{m}_{n,h_n}(x)$ and $\widehat{m}_{n,h_n}(x)\}$ equals*

$$\begin{aligned} \frac{2}{n\{h_n(x) + h_n(y)\}} &[\sigma_m^2(z_n(x,y))\kappa_2^{-1} \int K(v - \alpha_n(v))K(v + \alpha_n(v))\,dv \\ &+ \delta_n(x,y)f_X^{-2}(z_n(x,y))\{w_2^{(1)} - m^2 f_X^{(1)}\}(z_n(x,y)) \\ &\times \int vK(v - \alpha_n(v))K(v + \alpha_n(v))\,dv + o(\|h_n\|)]. \end{aligned}$$

Proof. The integral $EY^2K_{h_n(x)}(x-X)K_{h_n(y)}(y-X) = EY^2K_{h_n(x)}(x-X)K_{h_n(y)}(y-X) = \int K_{h_n(x)}(x-u)K_{h_n(y)}(y-u)w_2(u)\,du$ is expanded changing the variable u in v and it equals

$$\begin{aligned} &\frac{2}{h_n(x) + h_n(y)} \int K(v - \alpha_n(v))K(v + \alpha_n(v))w_2(u_n(x,y,v))\,dv \\ &= \frac{2}{h_n(x) + h_n(y)}\{w_2(z_n(x,y)) \int K(v - \alpha_n(v))K(v + \alpha_n(v))\,dv \\ &+ \delta_n(x,y)w_2^{(1)}(z_n(x,y)) \int vK(v - \alpha_n(v))K(v + \alpha_n(v))\,dv + o(\|h_n\|)\} \end{aligned}$$

then the L_2-approximation of $(nh_n(x))^{1/2}\{\widehat{m}_{n,h_n(x)} - m_{n,h_n(x)}\}(x)$ and Lemma 4.3 end the proof. □

Lemma 3.3 is extended to $\widehat{\mu}_{n,h_n}$ and $\widehat{m}_{n,h_n}$ with functional bandwidths like 4.2 and the weak convergence on $\mathcal{I}_{X,\|h_n\|}$ of the process with varying bandwidth $U_{n,h_n}(x) = (nh_n(x))^{1/2}\{\widehat{f}_{n,h_n(x)}(x) - f(x)\}$ is proved as for the density estimator.

Lemma 4.4. *For a regression function m and density f_X of class $C_s(\mathcal{I}_X)$, $s \geq 2$, and under Conditions 3.1 and 4.1, for every x and y in $\mathcal{I}_{X,h_n}$ the mean of the variation of $\widehat{m}_{n,h_n}$ between x and y has the order $O(|x-y|)$ and $E|\widehat{m}_{n,h_n}(x)-\widehat{m}_{n,h}(y)|^2 = O(n^{-1}\|h_n^{-1}\|^3(x-y)^2)$ if $|xh_n^{-1}(x)-yh_n^{-1}(y)| \leq 1$. Otherwise, it is a $O(n^{-1}\|h_n^{-1}\|)$.*

Theorem 4.2. *Under the conditions, for a density f of class $C_s(\mathcal{I}_X)$ and a varying bandwidth sequence such that $n\|h_n\|^{2s+1}$ converges to $\|h\|$, the process U_{n,h_n} converges weakly to the process defined on $\mathcal{I}_X$ as $W_m + h^{1/2}b_m$, where W_m is a continuous centered Gaussian process with covariance $\sigma_m^2(x)\delta_{\{x=x'\}}$ at x and x'.*

The estimators of the derivatives of the regression function are modified by the derivatives of the bandwidth and the kernel in each term of the estimators, as detailed in Appendix B, and the first derivative is $\widehat{m}_{n,h}^{(1)} = \widehat{f}_{n,h}^{-1}\{\widehat{\mu}_{n,h}^{(1)} - \widehat{m}_{n,h}\widehat{f}_{n,h}^{(1)}\}$, like in (3.15), with notations of the appendix for $d\{K_{h_n(x)}(x)\}/dx$. The results of Proposition 3.5 are extended to the estimator $\widehat{m}_{n,h_n}^{(k)}$ with a varying bandwidth sequence, its bias is a $O(\|h_n\|^s)$, and its variance a $O((n\|h_n^{-1}\|)^{2k+1})$, hence the optimal bandwidth is a $O(n^{-1/(2k+2s+1)})$ and the optimal mean squared error is a $O(n^{-2s/(2k+2s+1)})$.

In the regression model with a conditional variance function $\sigma^2(x)$, the kernel estimator (3.17) with continuous functional bandwidths h_n and δ_n can be written

$$\widehat{\sigma}^2_{n,h_n(x),\delta_n(x)}(x) = \frac{\sum_{i=1}^n \{Y_i - \widehat{m}_{n,h_n(x)}(X_i)\}^2 K_{\delta_n(x)}(x - X_i)}{\sum_{i=1}^n K_{\delta_n(x)}(x - X_i)},$$

then a new estimator for the regression function is defined using this estimator as a weighting process $\widehat{w}_n = \widehat{\sigma}^{-1}_{n,h_n,\delta_n}$ in the estimator of the regression function

$$\widehat{m}_{\widehat{w}_n,n,h_n(x)}(x) = \frac{\sum_{i=1}^n \widehat{w}_n(X_i) Y_i K_{h_n(x)}(x - X_i)}{\sum_{i=1}^n \widehat{w}_n(X_i) K_{h_n(x)}(x - X_i)}.$$

The bias and variance of the estimator $\widehat{\sigma}^2_{n,h_n(x),\delta_n(x)}(x)$ and the fixed bandwidth estimator for $\sigma^2(x)$ are still similar. The bias of $\widehat{m}_{\widehat{w}_n,n,h_n(x)}(x)$ and $\widehat{m}_{w,n,h_n(x)}(x)$ have the same approximations, the variance of $\widehat{m}_{w,n,h_n(x)}(x)$ is identical to the variance of $\widehat{m}_{n,h_n(x)}(x)$ whereas the variance of $\widehat{m}_{\widehat{w}_n,n,h_n(x)}(x)$ is modified like with the fixed bandwidth estimator. The weak convergence theorem 4.2 extends to the weighted regression estimator.

4.4 Estimation for processes

Let $(X_t)_{t\in[0,T]}$ be a continuously observed stationary and ergodic process satisfying (2.13), with values in $\mathcal{I}_X$. The limiting marginal density defined by (2.14) is estimated with an optimal bandwidth of order $O(T^{1/(2s+1)})$ as proved in Section 2.10. For every x in $\mathcal{I}_{X,T,\|h_T\|}$

$$\widehat{f}_{T,h_T(x)}(x) = \frac{1}{T}\int_0^T K_{h_T(x)}(X_s - x)\,ds \tag{4.3}$$

where $T^{1/(2s+1)}\|h_T\| = O(1)$. Conditions 2.1-2.2 are supposed to be satisfied, with a density f in class C_s and assuming that the bandwidth function fulfills Conditions 4.1 with the approximation

$$h_T(x) = T^{-1/(2s+1)}\{h^{1/(2s+1)}(x) + o(1)\}. \tag{4.4}$$

The results of the previous sections extends to prove that for every x in $\mathcal{I}_{X,T,\|h_T\|}$, the bias of $\widehat{f}_{T,h}(x)$ is $b_{T,h_T}(x) = \frac{h_T^s(x)}{s!}m_{sK}f^{(s)}(x) + o(\|h_T\|^s)$, its variance is

$$Var\{\widehat{f}_{T,h_T}(x)\} = (Th_T(x))^{-1}\kappa_2 f(x) + o((T^{-1}\|h_T^{-1}\|),$$

its covariances are $o((T^{-1}\|h_T^{-1}\|)$ and the L_p-norms are $\|\widehat{f}_{T,h_T}(x) - f_{T,h_T}(x)\|_p = 0((T^{-1}\|h_T^{-1}\|)^{1/p})$. The ergodic property (2.16) for k-dimensional vectors of values of the process $(X_t)_t$ entails the weak convergence of the finite dimensional distributions of the density estimator $\widehat{f}_{T,h}$. Lemma 4.2 extends to the ergodic process and entails the weak convergence of $(Th_T)^{1/2}(\widehat{f}_{T,h} - f)$ to a Gaussian process with variance $\kappa_2 f(x)$ at x and covariances zero.

For a continuously observed stationary and ergodic process $(X_t, Y_t)_{t\le T}$ with values in $\mathcal{I}_{X,Y}$, consider the regression model $Y_t = m(X_t) + \sigma(X_t)\varepsilon_t$ where $(\varepsilon_t)_{t\in[0,T]}$ is a Brownian motion such that $E(\varepsilon_t \mid X_t) = 0$ and $E(\varepsilon_t\varepsilon_s \mid X_t \wedge X_s) = E\{(\varepsilon_t \wedge \varepsilon_s)^2 \mid X_t \wedge X_s) = 1$. The bivariate process Z is supposed to be ergodic, satisfying the properties (2.13) and (2.16). Under the same conditions as in Chapter 3, the regression function m is estimated on an interval $\mathcal{I}_{X,Y,T,\|h_T\|}$ by the kernel estimator

$$\widehat{m}_{T,h_T}(x) = \frac{\int_0^T Y_s K_{h_T(x)}(x - X_s)\,ds}{\int_0^T K_{h_T(x)}(x - X_s)\,ds}.$$

The bias and variances established in Section 3.10 for the functions f and m of class C_s and fixed bandwidth h_T are modified, with the notation $\mu = mf$

$$\begin{aligned}
b_{m,T,h_T(x)}(x) &= h_T(x)^s b_m(x) + o(\|h_T\|^s),\\
b_m(x) &= \frac{m_{sK}}{s!} f_X^{-1}(x)\{\mu^{(s)}(x) - m(x) f_X^{(s)}(x)\},\\
v_{m,T,h}(x) &= (Th_T(x))^{-1}\sigma_m^2(x) + o((T\|h_T\|)^{-1}),\\
\sigma_m^2(x) &= \kappa_2 f_X^{-1}(x) Var(Y \mid X = x)
\end{aligned}$$

and the covariance of $\widehat{m}_{T,h_T(x)}(x)$ and $\widehat{m}_{T,h_T(x)}(y)$ is a $o((T\|h_T\|)^{-1})$. The weak convergence of the process $(Th_T(x))^{1/2}\{\widehat{m}_{T,hh_T(x)}(x) - m(x)\}$ is then proved by the same methods, under the ergodicity properties.

In a model with a variance function, the regression function is also estimated using a weighting process $\widehat{w}_T = \widehat{\sigma}_{T,h_T,\delta_T}^{-1}$ in the estimator of the regression function

$$\begin{aligned}
\widehat{\sigma}_{T,h_T,\delta_T}^2(x) &= \frac{\int_0^T \{Y_s - \widehat{m}_{T,h_T(X_s)}(X_s)\}^2 K_{\delta_T(x)}(x - X_s)\,ds}{\int_0^T K_{\delta_T(x)}(x - X_s)\,ds},\\
\widehat{m}_{\widehat{w}_T,T,h_T(x)}(x) &= \frac{\int_0^T \widehat{w}_T(X_i) Y_i K_{h_T(x)}(x - X_i)}{\int_0^T \widehat{w}_T(X_i) K_{h_T(x)}(x - X_i)}.
\end{aligned}$$

The previous modifications of the bias and variance of the estimator extend to the continuously observed process $(X_t)_{t \le T}$.

4.5 Exercises

(1) Compute the fixed and varying optimal bandwidths for the estimation of a density and compare the respective density estimators.
(2) Give the expressions of the first moments of the varying bandwidth estimator of the conditional probability $p(x) = P(Y|X = x)$ for a Y binary variable, conditionally on the value of a continuous variable X (Exercise 3.10-(2)).

(3) For the hierarchical observations of n independent sub-samples of J_i dependent observations of Exercise 2.11-(5), determine a varying bandwidth estimator for the limiting density f and ergodicity conditions for the calculus of its bias and variance, and write their first order approximations.

(4) Write the expressions of the bias and the variance of the continuous estimator $\widehat{F}_{Y|X,n,h_n(x)}$ for the distribution function of $Y \leq y$ conditionally on $X \leq x$ of Exercise 3.10-(8), with a varying bandwidth and prove its weak convergence.

Chapter 5

Nonparametric estimation of quantiles

5.1 Introduction

Let F be a distribution function with density f on $\mathbb{R}$, $\widehat{F}_n$ its empirical distribution function and $\nu_n = n^{1/2}(\widehat{F}_n - F)$ the normalized empirical process. The process $\widehat{F}_n - F$ convergences to zero uniformly a.s. and in L_2, and ν_n converges weakly to $B \circ F$, where B is the Brownian motion. The quantile $\widehat{Q}_n$ is the inverse functional for $\widehat{F}_n$, it converges therefore in probability to the inverse Q_F of F, uniformly on $[0, 1]$. The quantile estimator is approximated as

$$\widehat{Q}_n = Q_F - \{\nu_n \circ Q_F\}\{f \circ Q_F\}^{-1} + \{\nu_n \circ Q_F\}^2\{f' \circ Q_F\}\{f \circ Q_F\}^{-3} + o(\nu_n^2).$$

As a consequence, the quantile process

$$n^{1/2}(\widehat{Q}_n - Q_F) = -n^{1/2}\frac{\nu_n}{f} \circ Q_F + r_n$$

converges weakly to a centered Gaussian process with covariance function $\{F(s \wedge t) - F(s)F(t)\}\{f \circ Q_F(s)\}^{-1}\{f \circ Q_F(t)\}^{-1}$, for every s and t in $[0, 1]$. The remainder term is such that $\sup_{t \in [0,1]} \|r_n\|$ is a $o_{L_2}(1)$.

Consider the distribution function $F_{Y|X}$ of the variable Y conditionally on the regression variable X, in the model $Y = m(X) + \varepsilon$ with a continuous regression curve $m(x) = E(Y|X = x)$ and an observation error ε such that $E(\varepsilon|X) = 0$ and $Var(\varepsilon|X) = \sigma^2(X)$. It is defined with respect to the distribution function F_ε of ε by

$$F_{Y|X}(y; x) = P(Y \le y|X = x) = F_\varepsilon(y - m(x)). \tag{5.1}$$

The marginal distribution function of Y is $F_Y(y) = \int F_\varepsilon(y - m(s))\, dF_X(s)$ and the joint distribution function of (X, Y) is $F_{X,Y}(x, y) = \int 1_{\{s \le x\}} F_\varepsilon(y -$

$m(s))\,dF_X(s)$. The estimator of $F_{Y|X}$ is defined by smoothing the regression variable with a kernel is

$$\widehat{F}_{Y|X,n,h}(y;x)=\frac{\sum_{i=1}^n K_h(x-X_i)1_{\{Y_i\le y\}}}{\sum_{i=1}^n K_h(x-X_i)}$$

and an estimator of F_ε is deduced from those of $F_{Y|X}$, F_X and m as

$$\widehat{F}_{\varepsilon,n,h}(s)=n^{-1}\sum_{1\le i\le n}\widehat{F}_{Y|X,n,h}(s+\widehat{m}_{n,h}(X_i);X_i).$$

In this expression, the estimator of the regression function can be weighted by the inverse of the square root of the kernel estimator for the variance function σ^2. Therefore, all functions of the model, m, $\widehat{\sigma}^2$, $F_{Y|X}$ and F_ε, are easily estimated from the sample $(X_i,Y_i)_{i\le n}$.

The quantile of the conditional distribution function of Y given X are first defined with respect to Y, then with respect to X. For every t in $[0,1]$ and at fixed x in $\mathcal{I}_X$, the conditional distribution $F_{Y|X}(y;x)$ is increasing with respect to y and its inverse is defined as

$$Q_Y(t;x)=F_{Y|X}^{-1}(t;x)=\inf\{y\in\mathcal{I}_Y:F_{Y|X}(y;x)\ge t\}. \tag{5.2}$$

It is right-continuous with left-hand limits, like the $F_{Y|X}$. For every $x\in\mathcal{I}_X$, $F_{Y|X}\circ Q_Y(t;x)\ge t$ with equality if and only if $F_{Y|X}(x)$ is (x,y) belongs to the support of (X,Y). Assuming that the function m is monotone by intervals, the definition (5.1) implies the monotonicity on the same intervals of the conditional distribution function $F_{Y|X}$ with respect to the Y, with the inverse monotonicity.

On each interval of monotonicity and for every s in the image of $\mathcal{I}_X$, the quantile $Q_X(y;s)$ is defined by inversion of the conditional distribution $F_{Y|X}$ in the domain of the variable X, at fixed y, from equation (5.1)

$$Q_X(y;s)=\begin{cases}\inf\{x\in\mathcal{I}_X:F_{Y|X}(y;x)\ge t\}, & \text{if } m \text{ is decreasing,}\\ \sup\{x\in\mathcal{I}_X:F_{Y|X}(y;x)\le t\}, & \text{if } m \text{ is increasing.}\end{cases} \tag{5.3}$$

For every $y\in\mathcal{I}_Y$, $Q_X\circ F_{Y|X}(y;x)=x$ if and only if m and F_ε are continuous on $\mathcal{I}_Y$ and $F_{Y|X}\circ Q_X(y;s)=s$ if and only if m and F_ε are strictly monotone, for every (s,y) in $\mathcal{D}_{X,Y}$. The empirical conditional distribution function defines in the same way the empirical quantile processes $\widehat{Q}_{X,n,h}$ and $\widehat{Q}_{Y,n,h}$, according to (5.3) and (5.2) respectively. If (x,y) belongs to $\widehat{\mathcal{D}}_{X,Y,n,h}$, the marginal components x and y belong respectively to $\widehat{\mathcal{D}}_{X,n,h}$ and $\widehat{\mathcal{D}}_{Y,n,h}$ which are the domains of $\widehat{Q}_{X,n,h}$ and $\widehat{Q}_{Y,n,h}$, respectively.

Another question of interest for a regression function m monotone on an interval $\mathcal{I}_m$ is to determine its inverse with its distribution properties.

Consider a continuous regression function m, increasing on a sub-interval $\mathcal{I}_m$ of the support $\mathcal{I}_X$ of X, its inverse is defined as

$$m^{-1}(t) = \inf\{x \in \mathcal{I}_X : m(x) \geq t\}. \tag{5.4}$$

It is increasing and continuous on the image of $\mathcal{I}_m$ by m and satisfies $m^{-1} \circ m = m \circ m^{-1} = id$.

5.2 Asymptotics for the quantile processes

Let $\mathcal{I}_{X,Y,h} = \{(s,y) \in \mathcal{I}_{X,Y}; [s-h, s+h] \in \mathcal{I}_X\}$. Under conditions similar to those of the nonparametric regression, Proposition 3.1 applies considering y as fixed, with $f_X(x)F_{Y|X}(y;x) = \int 1_{\{\zeta \leq y\}} f_{X,Y}(x,\zeta)\, d\zeta$ instead of $\mu(x)$ and with the conditional function $F_{Y|X}(y;x)$, for every (x,y) in $\mathcal{I}_{X,Y}$. The weak convergence of the process defined on $\mathcal{I}_{X,h}$, at fixed y, by $(nh)^{1/2}\{\widehat{F}_{Y|X,n,h}(y;\cdot) - F_{Y|X}(y;\cdot)\}$ is a corollary of Theorem 3.1. The expressions of the bias and the L_p-norms rely on an expansion up to higher order terms of its moments.

Proposition 5.1. *Let F_{XY} be a distribution function of $C_{s+1}(I_{X,Y})$. Under Conditions 2.1 for the density f_X and 3.1 for the conditional distribution function $F_{Y|X}(y;x)$ at fixed y, the variable $\sup_{I_{X,Y,h}} |\widehat{F}_{Y|X,n,h} - F_{Y|X}|$ tends to zero a.s., its bias and its variance are*

$$b_{F_{Y|X},n,h}(y;x) = h^2 b_F(y;x) + o(h^2) \tag{5.5}$$

$$b_F(y;x) = \frac{1}{s!} m_{sK} f_X^{-1}(x)\{\frac{\partial^{s+1} F_{X,Y}(x,y)}{\partial x^{s+1}} - F_{Y|X}(x,y) f_X^{(2)}(x)\},$$

$$v_{F_{Y|X},n,h}(y;x) = (nh)^{-1} v_F(y;x) + o((nh)^{-1}) \tag{5.6}$$

$$v_F(y;x) = \kappa_2 f_X^{-1}(x) F_{Y|X}(y;x)\{1 - F_{Y|X}(y;x)\}.$$

At every fixed y in $\mathcal{I}_Y$, the process $(nh)^{1/2}\{\widehat{F}_{Y|X,n,h}(y) - F_{Y|X}(y)\}1_{\{\mathcal{I}_{X,h}\}}$ converges weakly to a Gaussian process defined in I_X, with mean function $\lim_n (nh^5)^{1/2} b_{F_{Y|X}}(y;\cdot)$, covariances zero and variance function $v_{F_{Y|X}}(y;\cdot)$.

The results for the bias of the estimator $\widehat{F}_{Y|X,n,h}$ extend to a density f_X in C_s as in Lemma 3.2. The weak convergence of the bivariate process $(nh)^{1/2}(\widehat{F}_{Y|X,n,h} - F_{Y|X})$ defined on $\mathcal{I}_{X,Y,h}$ requires an extension of the previous results as for the empirical distribution function of Y.

Proposition 5.2. *The process*

$$\nu_{Y|X,n,h} = (nh)^{1/2}\{\widehat{F}_{Y|X,n,h} - F_{Y|X})\}1_{\{\mathcal{I}_{X,Y,h}\}}$$

converges weakly to a Gaussian process W_ν on $I_{Y,X}$, with mean function $\lim_n (nh^5)^{1/2} b_F(y;\cdot)$, variance $v_{Y|X}$ and covariances at fixed x

$$Cov_{Y|X}(y, y'; x) = \kappa_2 f_X^{-1}(x)\{F_{Y|X}(y \wedge y'; x) - F_{Y|X}(y; x)F_{Y|X}(y'; x)\},$$

and zero otherwise.

Proof. This is a consequence of the weak convergence of the finite dimensional distributions of $\nu_{Y|X,n,h}$ and of its tightness, due to the bound obtained for the moments of the squared variations between (x, y) and (x', y') of the joint empirical process, $\nu_{Y|X,n,h}(y; x) - \nu_{Y|X,n,h}(y'; x) - \{\nu_{Y|X,n,h}(y; x') - \nu_{Y|X,n,h}(y'; x')\}$ is a $O((x'-x)^2 + (y'-y)^2)$. The bound $O((y'-y)^2)$ is obtained for the empirical process at fixed x and x', and $O((x'-x)^2)$ as in the proof of Lemma 3.3, at fixed y and y'. □

Let $F_{Y|X}(y; x)$ be monotone with respect to x. If n is sufficiently large, then $\widehat{F}_{Y|X,n,h}$ is monotone, as proved in the following lemma. The means are denoted $F_{Y|X,n,h}$, $Q_{Y,n,h}$ and $Q_{X,n,h}$ for $E\widehat{F}_{Y|X,n,h}$, $E\widehat{Q}_{Y,n,h}$ and, respectively, $E\widehat{Q}_{X,n,h}$.

Lemma 5.1. *If $n \geq n_0$ large enough, $F_{Y|X,n,h}$ is monotone on $\mathcal{I}_{X,Y,h}$. Moreover, if $F_{Y|X}$ is increasing with respect to x in $\mathcal{I}_X$ then, for every $x_1 < x_2$ and $\zeta > 0$, there exists $C > 0$ such that*

$$\Pr\{\widehat{F}_{Y|X,n,h}(x_2) - \widehat{F}_{Y|X,n,h}(x_1) > C\} \geq 1 - \zeta.$$

Proof. Let y be considered as fixed in $\mathcal{I}_Y$, $x_1 < x_2$ be in $\mathcal{I}_{X,h}$ and such that $F_{Y|X}(y; x_2) - F_{Y|X}(y; x_1) = d > 0$. For n large enough the bias of $\widehat{F}_{Y|X,n,h}(y; x_2) - \widehat{F}_{Y|X,n,h}(y; x_1)$ is strictly larger than $d/2$, by Proposition 5.2. The uniform consistency of Proposition 5.1 implies, for every η and $\zeta > 0$, the existence of an integer n_0 such that for every $n \geq n_0$, $\Pr\{|\widehat{F}_{Y|X,n,h}(y; x_1) - F_{Y|X}(y; x_1)| + |\widehat{F}_{Y|X,n,h}(y; x_2) - F_{Y|X}(y; x_2)| > \eta\} < \zeta$. For the monotonicity of the empirical conditional distribution function, let $d > \eta > 0$, then

$$\begin{aligned}
&\Pr\{\widehat{F}_{Y|X,n,h}(y; x_2) - \widehat{F}_{Y|X,n,h}(y; x_1) > d - \eta\} \\
&= 1 - \Pr\{(\widehat{F}_{Y|X,n,h} - F_{Y|X})(y; x_1) - (\widehat{F}_{Y|X,n,h} - F_{Y|X})(y; x_2) \geq \eta\} \\
&\geq 1 - \Pr\{|\widehat{F}_{Y|X,n,h} - F_{Y|X}|(y; x_2) + |\widehat{F}_{Y|X,n,h} - F_{Y|X}|(y; x_1) \geq \eta\} \\
&\geq 1 - \zeta.
\end{aligned}$$

□

The asymptotic behaviour of the quantile processes follows the same principles as the distribution functions. We first consider the quantile Q_Y defined by (5.2) conditionally on fixed $X = x$, it is always increasing. The empirical quantile function is increasing with probability tending to 1, as in Lemma 5.1 and the functions $Q_{X,n,h}$ and $Q_{Y,n,h}$ are monotone, for n large enough. The results of Section 5.1, are adapted to the empirical quantiles. Another quantile function is defined for n large enough by

$$\widetilde{Q}_{Y,n,h}(v;x) = \sup\{y : (x,y) \in \mathcal{I}_{X,Y,h}, F_{Y|X,n,h}(y;x) \le v\}, \; v \in \widehat{\mathcal{D}}_{Y,n,h}. \tag{5.7}$$

The uniform convergence of $F_{Y|X,n,h}$ to $F_{Y|X}$ implies that $\widetilde{Q}_{Y,n,h}$ converges uniformly to Q_Y. The derivative with respect to y of $F_{Y|X}(y;x)$ belonging to $C_2(\mathcal{I}_Y)$ is $f_{Y|X}(y;x)$, for every x in $\mathcal{I}_X$. Let b_F and v_F be defined by (5.6) and (5.6), respectively.

Proposition 5.3. *Let $F_{X|Y}$ be a continuous conditional distribution function, the process $\sup_{\widehat{\mathcal{D}}_{Y,n,h}\times\mathcal{I}_X} |\widehat{Q}_{Y,n,h} - Q_Y|(u;x)$ converges in probability to zero. If the density $f_{X,Y}$ of (X,Y) belongs to $C_s(\mathcal{I}_{X,Y})$, then for every x in $\mathcal{I}_X$ and u in $\widehat{\mathcal{D}}_{Y,n,h}$, the bias of $\widehat{Q}_{Y,n,h}$ equals*

$$b_Y(u;x) = -h^2 \frac{b_F}{f_{Y|X}} \circ Q_Y(u;x) + o(h^2),$$

and its variance is

$$v_Y(u;x) = (nh)^{-1} \frac{v_F}{\{f_{Y|X}\}^2} \circ Q_Y(u;x) + o((nh)^{-1}).$$

Proof. By definition of the inverse function, for every x in $\mathcal{I}_{X,n,h}$ and u in $\widehat{\mathcal{D}}_{Y,n,h}$, there exists an unique y in $\mathcal{I}_{Y,n,h}$ such that $u = \widehat{F}_{Y|X,n,h}(y;x)$, then by derivability of the inverse function

$$\begin{aligned}
\widehat{Q}_{Y,n,h}(u;x) - Q_Y(u;x) &= \widehat{Q}_{Y,n,h} \circ \widehat{F}_{Y|X,n,h}(y;x) - Q_Y \circ \widehat{F}_{Y|X,n,h}(x) \\
&= Q_Y \circ F_{Y|X}(y;x) - Q_Y \circ \widehat{F}_{Y|X,n,h}(y;x) \\
&= -\frac{\widehat{F}_{Y|X,n,h}(y;x) - F_{Y|X}(y;x)}{f_{Y|X}(y;x)} \\
&\quad + o(\widehat{F}_{Y|X,n,h}(y;x) - F_{Y|X}(y;x)).
\end{aligned}$$

The functions $\widetilde{Q}_{Y,n,h}$ satisfy a similar approximation with the functions $F_{Y|X,n,h}$. By the uniform convergence in probability of $F_{Y|X,n,h}$ to $F_{Y|X}$ on $\mathcal{I}_{X,n,h}$ and under the condition that the density is bounded away from zero on $\mathcal{I}_X$, the processes $\widehat{Q}_{Y,n,h}$ and the functions $\widetilde{Q}_{Y,n,h}$ convergence uniformly to Q_Y.

At fixed x, let u be in $\widehat{\mathcal{D}}_{Y,n,h}$. In order to calculate the bias and variance of $\widehat{Q}_{Y,n,h}(u)$, we first determine the order of the bias and the variance of the processes

$$\widehat{\eta}_{Y,n,h}(u) = F_{Y|X} \circ \widehat{Q}_{Y,n,h}(u) - F_{Y|X} \circ \widetilde{Q}_{Y,n,h}(u) \tag{5.8}$$

which converge in probability to zero. Then the quantile estimator satisfies

$$\widehat{Q}_{Y,n,h} = Q_Y \circ (\widehat{\eta}_{Y,n,h} + F_{Y|X} \circ \widetilde{Q}_{Y,n,h}). \tag{5.9}$$

Taylor expansions allow to express $\widetilde{Q}_{Y,n,h}$ as a function of Q and $\widehat{Q}_{Y,n,h}$ as a function of $\widetilde{Q}_{Y,n,h}$ and of the process $\widehat{\eta}_{Y,n,h}$. Since $\widehat{F}_{Y|X,n,h} \circ \widehat{Q}_{Y,n,h}$ and $F_{Y|X,n,h} \circ \widetilde{Q}_{Y,n,h}$ equal identity, (5.8) is also written

$$\widehat{\eta}_{Y,n,h}(u) = b_{F,n,h} \circ \widetilde{Q}_{Y,n,h}(u) - \{\widehat{F}_{Y|X,n,h} - F_{Y|X}\} \circ \widehat{Q}_{Y,n,h}(u).$$

$$E\{\widehat{\eta}_{Y,n,h}(u)\} = h^2 b_F \circ \widetilde{Q}_{Y,n,h}(u) - h^2 E\{b_F \circ \widehat{Q}_{Y,n,h}(u)\} + o(h^2), \tag{5.10}$$

and the variance of $\widehat{\eta}_{Y,n,h}(u)$ equals

$$\begin{aligned} Var\{\widehat{\eta}_{Y,n,h}(u)\} &= E\left[Var\{\widehat{F}_{Y|X,n,h} \circ \widehat{Q}_{Y,n,h}(u) | \widehat{Q}_{Y,n,h}(u)\}\right] \\ &\quad + Var\{b_{F,n,h} \circ \widehat{Q}_{Y,n,h}(u)\} \\ &= (nh)^{-1} E\{v_F \circ \widehat{Q}_{Y,n,h}(u)\} \\ &\quad + h^4 Var\{b_F \circ \widehat{Q}_{Y,n,h}(u)\} + o(n^{-1}h^{-1} + h^4). \end{aligned} \tag{5.11}$$

The moments of order $l \geq 3$ of $\widehat{\eta}_{Y,n,h}$ are bounded using an expansion of the moments of the sum in Equation (5.8) by

$$2^l \left\{ |b_{Y,n,h} \circ \widetilde{Q}_{Y,n,h}(u)|^l + E|(\widehat{F}_{Y|X,n,h} - F_{Y|X}) \circ \widehat{Q}_{Y,n,h}(u)|^l \right\},$$

the second right hand term $E\{|(\widehat{F}_{Y|X,n,h} - F_{Y|X,n,h} + b_{Y,n,h}) \circ \widehat{Q}_{Y,n,h}(u)|^l\}$ is lower than

$$\begin{aligned} 2^l E[E\{|(\widehat{F}_{Y|X,n,h} - F_{Y|X,n,h}) \circ \widehat{Q}_{Y,n,h}(u)|^l | \widehat{Q}_{Y,n,h}(u)\} \\ + |b_{Y,n,h} \circ \widehat{Q}_{Y,n,h}(u)|^l], \end{aligned}$$

thus

$$\begin{aligned} E|\widehat{\eta}_{Y,n,h}(u)|^l \leq 2^l \Big[|b_{Y,n,h} \circ \widetilde{Q}_{Y,n,h}(u)|^l + 2^l E\Big\{ |b_{Y,n,h} \circ \widehat{Q}_{Y,n,h}(u)|^l \\ + E\{|(\widehat{F}_{Y|X,n,h} - F_{Y|X,n,h}) \circ \widehat{Q}_{Y,n,h}(u)|^l | \widehat{Q}_{Y,n,h}(u)\} \Big\} \Big]. \end{aligned}$$

By Propositions 2.2 and 3.1, the conditional expectation $E\{|\widehat{F}_{Y|X,n,h} - F_{Y|X,n,h}|^l$ is $O((nh)^{-l/2})$, and both terms in $b_{Y,n,h}$ are $O(h^{2l})$, hence $E|\widehat{\eta}_{n,h}(u)|^l = o((nh)^{-1})$ for every $l \geq 3$. The expression of the bias of

$\widehat{F}_{Y|X,n,h}$ implies $\{F_{Y|X,n,h} - F_{Y|X}\} \circ \widetilde{Q}_{n,h}(u) = h^2 b_F \circ \widetilde{Q}_{n,h}(u) + o(h^2)$, therefore

$$\begin{aligned}\widetilde{Q}_{Y,n,h}(u) &= F^{-1}_{Y|X}(u - h^2 b_F \circ \widetilde{Q}_{Y,n,h}(u)) + o(h^2)\\ &= Q_Y(u) - h^2 \frac{b_F \circ \widetilde{Q}_{Y,n,h}(u)}{f_{Y|X} \circ Q_Y(u)} + o(h^2).\end{aligned}$$

Since $\widetilde{Q}_{Y,n,h}(u) = Q_Y(u) + O(h^2)$, $b_F \circ \widetilde{Q}_{Y,n,h}(u) = b_F \circ Q_Y(u) + O(h^2)$, so that

$$\widetilde{Q}_{Y,n,h}(u) = Q_Y(u) - h^2 \frac{b_F}{f_{Y|X}} \circ Q_Y(u) + o(h^2). \tag{5.12}$$

Furthermore, by (5.8),

$$\begin{aligned}\widehat{Q}_{Y,n,h}(u) &= F^{-1}_{Y|X}(F_{Y|X} \circ \widetilde{Q}_{Y,n,h}(u) + \widehat{\eta}_{Y,n,h}(u))\\ &= \widetilde{Q}_{Y,n,h}(u) + \frac{\widehat{\eta}_{Y,n,h}(u)}{f_{Y|X} \circ \widetilde{Q}_{Y,n,h}(u)} + O(\widehat{\eta}^2_{Y,n,h}(u)),\end{aligned}$$

and, using (5.12),

$$\begin{aligned}\widehat{Q}_{Y,n,h}(u) = \widetilde{Q}_{Y,n,h}(u) &+ \frac{\widehat{\eta}_{Y,n,h}(u)}{f_{Y|X} \circ Q_Y(u)}\\ &+ O(h^2 \widehat{\eta}_{Y,n,h}(u)) + O(\widehat{\eta}^2_{Y,n,h}(u)).\end{aligned} \tag{5.13}$$

The expansion (5.12) implies $b_F \circ \widetilde{Q}_{Y,n,h}(u) = b_F \circ Q_Y(u) + o(1)$. With (5.13) and since $E\{\widehat{\eta}_{Y,n,h}(u)\}$ and $Var\{\widehat{\eta}_{Y,n,h}(u)\}$ are $o(1)$

$$\begin{aligned}b_F \circ \widehat{Q}_{Y,n,h}(u) &= b_F \circ \widetilde{Q}_{Y,n,h}(u) + \frac{\widehat{\eta}_{Y,n,h}(u)}{F_{Y|X} \circ Q_Y(u)} b_f \circ \widetilde{Q}_{Y,n,h}(u)\\ &\quad + O(h^2 \widehat{\eta}_{Y,n,h}(u)) + O(\widehat{\eta}^2_{Y,n,h}(u)),\\ E\{b_F \circ \widehat{Q}_{Y,n,h}(u)\} &= b_F \circ \widetilde{Q}_{Y,n,h}(u) + o(1) = b_F \circ Q_Y(u) + o(1).\end{aligned}$$

Moreover, $Var\{b_F \circ \widehat{Q}_{Y,n,h}(u)\} = O(h^4 + n^{-1}h^{-1})$ because of the approximations $Var\{\widehat{\eta}_{Y,n,h}(u)\} = O(h^4 + n^{-1}h^{-1})$ and $E\{\widehat{\eta}^4_{Y,n,h}(u)\} = o(n^{-1}h^{-1})$. From (5.10), the expectation of $\widehat{\eta}_{Y,n,h}(u)$ becomes

$$E\{\widehat{\eta}_{Y,n,h}(u)\} = o(h^2). \tag{5.14}$$

In the expansion (5.11), $E\{v_F \circ \widehat{Q}_{Y,n,h}(u)\} = v_F \circ Q_Y(u) + o(1)$ and $h^4 Var\{b_F \circ \widehat{Q}_{Y,n,h}(u)\} = O(h^8 + n^{-1}h^3) = o(n^{-1}h^{-1})$. The variance of $\widehat{\eta}_{Y,n,h}(u)$ is then equal to

$$Var\{\widehat{\eta}_{Y,n,h}(u)\} = (nh)^{-1} v_F \circ Q_Y(u) + o((nh)^{-1}). \tag{5.15}$$

Finally, the bias of $\widehat{Q}_{Y,n,h}(u)$ is deduced from (5.9), (5.12), (5.13) and (5.14), which imply

$$\widehat{Q}_{Y,n,h} = Q_Y + \{(\widehat{\eta}_{Y,n,h} + F_{Y|X} \circ \widetilde{Q}_{Y,n,h} - F_{Y|X} \circ Q_Y)/(f_{Y|X} \circ Q_Y)\}\{1 + o(1)\}$$

therefore

$$E\{\widehat{Q}_{Y,n,h}(u)\} = Q_Y(u) - h^2 \frac{b_F}{f_{Y|X} \circ Q_Y(u)} + o(h^2).$$

Equations (5.13) and (5.15) yield the variance of $\widehat{Q}_{Y,n,h}(u)$. □

Theorem 5.1. *The process $U_{Y,n,h} = (nh)^{1/2}\{\widehat{Q}_{Y,n,h} - Q_Y\}1_{\{\widehat{\mathcal{D}}_{Y,n,h}\}}$ converges weakly to $U_Y = \dfrac{W_\nu + \gamma^{1/2} b_F}{f_{Y|X}} \circ Q_Y$ where W_ν is the Gaussian process limit of $\nu_{Y|X,n,h}$.*

Proof. For every x in $\mathcal{I}_{X,n,h}$ and for every u in $\widehat{\mathcal{D}}_{Y,n,h}$, there exists an unique y in $\mathcal{I}_{Y,n,h}$ such that $u = \widehat{F}_{Y|X,n,h}(y;x)$ therefore

$$\begin{aligned}(\widehat{Q}_{Y,n,h} - Q_Y)(u;x) &= (\widehat{Q}_{Y,n,h} \circ \widehat{F}_{Y|X,n,h} - Q_Y \circ \widehat{F}_{Y|X,n,h})(y;x)\\ &= (Q_Y \circ F_{Y|X} - Q_Y \circ \widehat{F}_{Y|X,n,h})(y;x).\end{aligned}$$

From the convergence of $\widehat{F}_{Y|X,n,h}$ to $F_{Y|X}$, it follows

$$(nh)^{1/2}\{\widehat{Q}_{Y,n,h} - Q_Y\} = -\frac{\{\nu_{Y|X,n,h} + \gamma^{1/2} b_Y\}}{f_{Y|X,n,h}} \circ \widehat{Q}_{Y,n,h},$$

and its limit is deduced from Proposition 5.2. □

The representation of the conditional quantile process

$$\widehat{Q}_{Y,n,h} = Q_Y + \{(\widehat{\eta}_{Y,n,h} + F_{Y|X} \circ \widetilde{Q}_{Y,n,h} - F_{Y|X} \circ Q_Y)/(f_{Y|X} \circ \widehat{Q}_{Y,n,h})\} + r_{Y,n,h}, \tag{5.16}$$

where $\widehat{\eta}_{Y,n,h}$ is defined by (5.8) and where the remainder term $r_{Y,n,h}$ is $o_{L_2}((nh_n)^{-1/2})$, was established to prove Proposition 5.3 and Theorem 5.1. An analogous representation holds for the quantile process $\widehat{Q}_{X,n,h}$

$$\widehat{Q}_{X,n,h} = Q_X + \{(\widehat{\zeta}_{X,n,h} + F_{Y|X} \circ \widetilde{Q}_{X,n,h} - F_{Y|X} \circ Q_X)/(f_{Y|X} \circ Q_X)\} + r_{X,n,h}$$

where $\widehat{\zeta}_{X,n,h} = F_{Y|X} \circ \widehat{Q}_{X,n,h} - F_{Y|X} \circ \widetilde{Q}_{X,n,h}$ and $r_{X,n,h} = o_{L_2}((nh_n)^{-1/2})$. The bias b_X, the variance v_X and the weak convergence of $\widehat{Q}_{X,n,h}$ are deduced. Let $F^{(1)}_{Y|X}$ be the derivative of $F_{Y|X}(y;x)$ with respect to x.

Proposition 5.4. *Let $F_{X|Y}$ be a continuous conditional distribution function, the process $\sup_{\widehat{\mathcal{D}}_{X,n,h} \times \mathcal{I}_Y} |\widehat{Q}_{X,n,h} - Q_X|(u;x)$ converges in probability* *to zero. If the density $f_{X,Y}$ of (X,Y) belongs to $C_s(\mathcal{I}_{X,Y})$, then for every x in $\mathcal{I}_X$ and u in $\widehat{\mathcal{D}}_{X,n,h}$, the bias of $\widehat{Q}_{X,n,h}$ equals*

$$b_X(y;u) = -h^2 \frac{b_F}{\partial F_{Y|X}/\partial x} \circ Q_X(y;u) + o(h^2),$$

and its variance is

$$v_X(y;u) = (nh)^{-1} \frac{v_F}{\{\partial F_{Y|X}/\partial x\}^2} \circ Q_X(y;u) + o((nh)^{-1}).$$

Theorem 5.2. *The process $U_{X,n,h} = (nh)^{1/2}\{\widehat{Q}_{X,n,h} - Q_X\}1_{\{\widehat{\mathcal{D}}_{X,n,h}\}}$ converges weakly to $U_X = \dfrac{W_\nu + \lim_n (nh_n^5)^{1/2} b_F}{\partial F_{Y|X}/\partial x} \circ Q_X$.*

5.3 Bandwidth selection

The error criteria measuring the accuracy of $\widehat{Q}_{Y,n,h}(u)$ as an estimator of $Q(u)$ are generally sums of the variance and the squared bias of $\widehat{Q}_{Y,n,h}(u)$, where the variance increases as h tends to zero whereas the bias decreases. Under the assumption that $F_{Y|X}$ is twice continuously differentiable with respect to y and using results of Proposition 5.3, the mean squared error $MSE_Y(h) = E\left\{\widehat{Q}_{Y,n,h}(u) - Q(u)\right\}^2$ is asymptotically equivalent to

$$\begin{aligned}\mathrm{AMSE}_{Q_Y}(u;x,h) &= (nh)^{-1} \frac{v_F \circ Q_Y(u;x)}{\{f_{Y|X} \circ Q_Y(u;x)\}^2} \\ &\quad + h^4 \left\{ \frac{b_F}{f_{Y|X}} \circ Q_Y(u;x) \right\}^2 .\end{aligned}$$

Its minimization in h leads to an optimal local bandwidth, varying with u and x

$$h_{opt,loc}(u;x) = n^{-1/5} \left\{ \frac{v_F \circ Q_Y(u;x)}{4b_F^2 \circ Q_Y(u;x)} \right\}^{1/5} .$$

That this also the optimal local bandwidth minimizing the AMSE of $\widehat{F}_{Y|X,n,h}(u;x)$ for the unique value of x such that $y = Q_Y(u)$, that is

$$\mathrm{AMSE}_F(u;x,h) = (nh)^{-1} v_F(u;x) + h^4 b_F^2(u;x).$$

If the density f_X has a continuous derivative, that is also identical to the optimal local bandwidth minimizing the AMSE of $\widehat{Q}_{X,n,h}(y;x)$, at fixed y, by Proposition 5.4. Since the optimal rate for the bandwidth has the order $n^{-1/5}$, the optimal rate of convergence of $\widehat{Q}_{Y,n,h}$ to Q_Y is $n^{-4/5}$ and the process $(nh)^{1/2}\{\widehat{Q}_{Y,n,h}-Q_Y\}$ converges to a non-centered Gaussian process with an expectation different from zero because $nh^5 = O(1)$. Estimating the bias b_F and the variance v_F by bootstrap allows to estimate the optimal bandwidths for the quantile estimator without knowledge of its order s of derivability. A direct kernel estimator of the variance function of the process $\nu_{Y|X,n,h}$ is $\widehat{v}_{F,n,h} = \kappa_2 \widehat{f}_{X,n,h}^{-1} \widehat{F}_{|X,n,h}(1-\widehat{F}_{|X,n,h})$, according to the expression (5.6) of v_F.

For a conditional distribution function $F_{Y|X}(\cdot;x)$ having derivatives of order s, the bias is modified

$$\begin{aligned} b_{s,F,n,h}(y;x) &= h^s b_F(y;x) + o(h^s) \\ &= \frac{h^s}{s!} m_{sK} f_X^{-1}(x)\{\frac{\partial^{s+1} F_{X,Y}(x,y)}{\partial x^{s+1}} \\ &\qquad - F_{Y|X}(x,y) f_X^{(s)}(x)\} + o(h^2), \end{aligned}$$

and the optimal local bandwidth is modified by this s-order bias.

The global mean integrated squared error criteria are defined by integrating over all values of u in $\widehat{\mathcal{D}}_{Y,n,h}$ the $AMSE_{Q_Y}(u;x,h)$, conditionally on a fixed value of x

$$\begin{aligned} \text{AMISE}_{Q_Y}(x,h) &= \int E\{\widehat{Q}_{Y,n,h}(u;x) - Q_Y(u;x)\}^2 du \\ &= \int [(nh)^{-1} \frac{v_F \circ Q_Y(u;x)}{\{f_{Y|X} \circ Q_Y(u;x)\}^2} + h^4 \{\frac{b_Y \circ Q_Y(u;x)}{f_{Y|X} \circ Q_Y(u;x)}\}^2]\, du \\ &= \int_{\mathcal{I}_{Y,n,h}} \frac{\text{AMSE}_F(y;x,h)}{f_{Y|X}(y;x)}\, dy, \end{aligned}$$

which differs from the integral $\text{AMISE}_F(x,h) = \int_{\mathcal{I}_{Y,n,h}} \text{AMSE}_F(y;x,h)\, dy$, conditional to $X = x$. The mean of the conditional random criterion $\text{AMISE}_{Q_Y}(X,h)$ is

$$\int_{\mathcal{I}_{X,Y,n,h}} \frac{\text{AMSE}_F(y;x,h)}{f_{Y|X}(y;x)}\, dy\, dF_X(x).$$

In the same way, the global mean integrated squared error criteria is defined by integrating $\text{AMSE}_{Q_X}(y,h)$ over $\widehat{\mathcal{D}}_{X,n,h}$, for a fixed value of y, $\text{AMISE}_{Q_X}(y,h) = \int_{\mathcal{I}_{X,n,h}} \text{AMSE}_F(y;x,h)\{f_{Y|X}(y;x)\}^{-1}\, dx$, at fixed y. The global AMISE criteria for Q_X and Q_Y defined as integrals over $\widehat{\mathcal{D}}_{X,Y,n,h}$ are both equal to

$$\text{AMISE}_Q(h) = \int_{\mathcal{I}_{X,Y,n,h}} \frac{\text{AMSE}_F(y;x,h)}{f_{Y|X}(y;x)}\, dx\, dy$$

and they differ from the global criterion

$$\mathrm{AMISE}_F = \int_{\mathcal{I}_{X,Y,n,h}} \mathrm{AMSE}_F(y; x, h)\, dx\, dy = E \frac{\mathrm{AMSE}_F(Y; X, h)}{f_{X,Y}(X, Y)}.$$

Some discretized versions of these criteria are the Asymptotic Mean Average Squared Errors such as the AMASE_F corresponding to AMISE_F, $\mathrm{AMASE}_{Q_Y}(x, h)$ corresponding to $\mathrm{AMISE}_{Q_Y}(x, h)$ and $E\mathrm{AMISE}_{Q_Y}(X, h)$ are respectively defined by

$$\mathrm{AMASE}_F = n^{-1} \sum_{i=1}^{n} \frac{\mathrm{AMSE}_F(Y_i; X_i)}{f_{X,Y}(X_i, Y_i, h)},$$

$$\mathrm{AMASE}_{Q_Y}(x, h) = n^{-1} \sum_{i=1}^{n} \frac{\mathrm{AMSE}_F(Y_i; x, h)}{f_{Y|X}^2(Y_i; x, h)},$$

$$\mathrm{AMASE}_{Q_Y}(h) = n^{-2} \sum_{i=1}^{n} \sum_{j=1}^{n} \frac{\mathrm{AMSE}_F(Y_i; X_j, h)}{f_{Y|X}^2(Y_i; X_j, h)},$$

which is the empirical mean of $\mathrm{AMASE}_{Q_Y}(X, h)$. Similar ones are defined for Q_X and other means. Note that no computation of the global errors and bandwidths require the computation of integrals of errors for the empirical inverse functions, all are expressed through integrals or empirical means of AMSE_F with various weights depending on the density of X and the conditional density of Y given X. The optimal window for $\mathrm{AMASE}_{Q_F}(h)$ is

$$n^{-1/5} \left[\frac{\sum_{i=1}^n \{v_F (f_{X,Y})^{-1}\}(X_i, Y_i, h)}{4 \sum_{i=1}^n \{b_F^2 (f_{X,Y})^{-1}\}(X_i, Y_i, h)} \right]^{1/5},$$

for $\mathrm{AMASE}_{Q_Y}(h)$ it is

$$n^{-1/5} \left[\frac{\sum_{i=1}^n \sum_{j=1}^n \{v_F (f_{X,Y})^{-2}\}(X_i, Y_j, h)}{4 \sum_{i=1}^n \sum_{j=1}^n \{b_F^2 (f_{X,Y})^{-1}\}(X_i, Y_j, h)} \right]^{1/5}.$$

The expressions of other optimal global bandwidths are easily written and all are estimated by plugging estimators of the density, the bias b_F and the variance v_F with another bandwidth. The derivatives of the conditional distribution function are simply the derivatives of the conditional empirical distribution function, as nonparametric regression curves. The mean squared errors and the optimal bandwidths for the quantile process $\widehat{Q}_{X,n,h}$ are written in similar forms, with the bias b_X and variance v_X.

5.4 Estimation of the conditional density of Y given X

The conditional density $f_{Y|X}(y;x)$ is deduced from the conditional distribution function $F_{Y|X}(y;x)$ by derivative with respect to y and it is estimated using the kernel K with another bandwidth h'

$$\begin{aligned}\widehat{f}_{Y|X,n,h,h'}(y;x) &= \int K_{h'}(v-y)\,\widehat{F}_{Y|X,n,h}(dv;x)\\ &= \sum_{j=1}^{n} K_{h'}(Y_j-y)\frac{\sum_{i=1}^{n} K_h(x-X_i)I_{\{Y_i\le Y_j\}}}{\sum_{i=1}^{n} K_h(x-X_i)}.\end{aligned}$$

Proposition 5.5. *If the conditional density $f_{Y|X}$ belongs to the class $C_s(\mathcal{I}_{XY})$, $s\ge 2$, the process $(nhh')^{1/2}(\widehat{f}_{Y|X,n,h,h'}-f_{Y|X})$ converges weakly to a Gaussian process with mean $\lim_n(nh_nh'_n)^{1/2}(E\widehat{f}_{Y|X,n,h,h'}-f_{Y|X})$, with covariances zero and variance function $v_f=\kappa_2 f_{Y|X}(1-f_{Y|X})$.*

If $h'_n=h_n$ and $s=2$, the process $n^{1/2}h_n(\widehat{f}_{Y|X,n,h_n}-f_{Y|X})$ converges weakly to a Gaussian process with mean $\lim_n(nh_n^6)^{1/2}b_f$ where

$$b_f=\frac{1}{2}m_{2K}f_X^{-1}\{\partial^2 f_{X,Y}/\partial y^2+\partial^2 f_{X,Y}/\partial x^2-f_{Y|X}f_X^{(2)}\},$$

with covariances zero and variance function v_f. The optimal bandwidth is $O(n^{-1/6})$.

Proof. By Proposition 5.1, if $f_{Y|X}$ belongs to $C_2(\mathcal{I}_{XY})$, its expectation develops as

$$\begin{aligned}E\widehat{f}_{Y|X,n,h,h'}(y;x) &= \int K_{h'}(v-y)\,E\widehat{F}_{Y|X,n,h}(dv;x)\\ &= \int K_{h'}(v-y)\,(F_{Y|X}+b_{F,n,h})(dv;x)\\ &= f_{Y|X}(y;x)+\frac{h'^2}{2}m_{2K}\frac{\partial^2 f_{Y|X}(y;x)}{\partial y^2}\\ &\quad+\frac{\partial b_{F,n,h}(y;x)}{\partial y}+o(h^2)+o(h'^2).\end{aligned}$$

Assuming that $h'=h$, its bias $b_{f,n,h}(y;x)=h^2b_f(y;x)+o(h^2)$, with

$$b_f=\frac{1}{2}m_{2K}f_X^{-1}\{\partial^2 f_{X,Y}/\partial y^2+\partial^2 f_{X,Y}/\partial x^2-f_{Y|X}f_X^{(2)}\}.$$

Generally, the range of the variables X and Y differs and two distinct kernels have to be used, the bias is then expressed as a sum of two terms

$$b_{f,n,h,h'}(y;x) = h^2 b_F^{(1)}(y;x) + h'^2 b_f(y;x) + o(h^2) + o(h'^2),$$
$$b_F^{(1)} = \frac{1}{2} m_{2K} f_X^{-1}\{\partial^2 f_{X,Y}/\partial x^2 - f_{Y|X} f_X^{(2)}\},$$
$$b_f = \frac{1}{2} m_{2K} f_X^{-1} \partial^2 f_{X,Y}/\partial y^2.$$

The variance of the estimator is the limit of $Var\widehat{f}_{Y|X,n,h,h'}(y;x)$ written

$$\int K_{h'}(u-y)K_{h'}(v-y)Cov\{\widehat{F}_{Y|X,n,h}(du;x), \widehat{F}_{Y|X,n,h}(dv;x)\}$$
$$= (nh)^{-1}\kappa_2 f_X^{-1}(x)\{\int K_{h'}^2(u-y)\, F_{Y|X}(du;x)$$
$$- \int K_{h'}(u-y)K_{h'}(v-y)\, F_{Y|X}(du;x)F_{Y|X}(dv;x)\}.$$

The first integral develops as $I_1 = h'^{-1}\{\kappa_2 f_{Y|X}(y;x) + o(1)\}$, the second integral $I_2 = \int K_{h'}(u-y)K_{h'}(v-y)\, F_{Y|X}(du;x)F_{Y|X}(dv;x)$ is the sum of the integral outside the diagonal $\mathcal{D}_Y = \{(u,v) \in \mathcal{I}_{X,T}^2; |u-v| < 2h'_n\}$, which is zero, and an integral restricted to the diagonal which is expanded by changing the variables like in the proof of Proposition 2.2.

Let $\alpha_{h'}(u,v) = |u-v|/(2h')$, $u = y + h'(z + \alpha_{h'})$, $v = y + h'(z - \alpha_{h'})$ and $z = \{(u+v)/2 - y\}/(h')$

$$I_2 = \int_{\mathcal{D}_Y} K_{h'}(u-x)K_{h'}(v-x) f_{Y|X}(u;x) f_{Y|X}(v;x)\, du\, dv$$
$$= h'^{-1}\{\int_{\mathcal{D}_Y} K(z - \alpha_{h'}(u,v))K(z + \alpha_{h'}(u,v))\, dz du f_{Y|X}^2(y;x) + o(1)\}$$

and it is equivalent to $h'^{-1}\kappa_2 f_{Y|X}^2(y;x) + o(h'^{-1})$. The variance of the estimator of the conditional density $f_{Y|X}$ is then

$$v_{f_{Y|X,n,h,h'}} = (nhh')^{-1} v_f(y;x) + o((nhh')^{-1}),$$
$$v_f(y;x) = \kappa_2 f_{Y|X}(1 - f_{Y|X}) \tag{5.17}$$

and its covariances at every $y \neq y'$ tends to zero. □

The asymptotic mean squared error for the estimator of the conditional density is $MSE_{f_{Y|X}} y;x;h_n,h'_n) = h_n^4 b_F^{(1)2}(y;x) + h_n'^4 b_f^2(y;x) + (nh_n h'_n)^{-1} v_f$, it is minimal at the optimal bandwidth

$$h'_{n,opt,f_{Y|X}}(y;x) = \left\{\frac{1}{nh_n}\frac{v_f}{4b_f^2}(y;x)\right\}^{1/5}.$$

In this expression, h_n can be chosen as the optimal bandwidth for the kernel estimator of the conditional distribution function $F_{Y|X}$. The convergence rate $(nh_n h'_n)^{1/2}$ of the estimator for the conditional density is smaller than the convergence rate of a density and than $(nh_n^2)^{1/2} = O(n^{2/5})$, at the optimal bandwidth.

Assuming that $h'_n = h_n$, the optimal bandwidth is now

$$h_{n,opt,f_{Y|X}}(y;x) = \left\{ \frac{1}{nh} \frac{v_f}{2b_f^2}(y;x) \right\}^{1/6}$$

and the convergence rate for the estimator of the conditional density $f_{Y|X}$ is $n^{1/3}$ which is larger than the previous rate with two optimal bandwidths.

The mode of the conditional density $f_{Y|X}$ is estimated by the mode of its estimator and the proof of Proposition 2.5 applies with the modified rates of convergence and limit. The derivative of $\widehat{f}_{Y|X,n,h}(y;x)$ with respect to y converges with the rate $(nh_n^{'3} h_n)^{1/2}$ that is $n^{1/2}h_n^2$ for identical bandwidths.

5.5 Estimation of conditional quantiles for processes

Let $(Z_t)_{t\in[0,T]} = (X_t, Y_t)_{t\in[0,T]}$ be a continuously observed stationary and ergodic process with values in a metric space $\mathcal{I}_{XY}$ and the regression model $Y_t = m(X_t) + \sigma(X_t)\varepsilon_t$ as in Section 3.10. Under the ergodicity condition (2.13) for $(Z_t)_{t>0}$, the conditional distribution function of the limiting distribution corresponds to $F_{Y|X}(y;x)$ for a sample of variables and it is estimated from the sample-path of the process on $[0,T]$, similarly to (3.22), with a bandwidth indexed by T

$$\widehat{F}_{Y|X,T,h_T}(y;x) = \frac{\int_0^T 1_{\{Y_s \le y\}} K_{h_T}(x - X_s)\, ds}{\int_0^T K_{h_T}(x - X_s)\, ds}. \tag{5.18}$$

The numerator of (5.18), $\widehat{\mu}_{F,T,h_T}(y;x) = \frac{1}{T}\int_0^T 1_{\{Y_s \le y\}} K_{h_T}(x - X_s)\, ds$, has the mean $\mu_{F,T,h_T}(x) = \int_{\mathcal{I}_X} K_{h_T}(x-u)\, F_{X_s,Y_s}(du, y) = F(y;x)f(x) + h_T^2 b_F(y;x) + o(h_T^2)$ with $b_\mu(y;x) = \partial^3\{F(x,y)\}/\partial x^3$, for a conditional density of $C_2(\mathcal{I}_{XY})$.

Proposition 5.6. *Under the ergodicity conditions and for a conditional density $f_{Y|X}$ in class $C_s(\mathcal{I}_{XY})$, the bias b_{F,T,h_T} and the variance v_{F,T,h_T}*

of the estimator $\widehat{F}_{Y|X,T,h_T}$ *are*

$$\begin{aligned} b_{F,T,h_T}(y;x) &= h_T^s b_F(y;x) + o(h_T^s), \\ b_F(y;x) &= \frac{m_{sK}}{s!} f_X^{-1}(x)\{\partial^{s+1}F(x,y)/\partial x^{s+1} - F(x)f_X^{(s)}(x)\}, \\ v_{F,T,h_T}(y;x) &= (Th_T)^{-1}\{\sigma_F^2(y;x) + o(1)\}, \\ \sigma_F^2(x) &= \kappa_2 f_X^{-1}(x) F_{Y|X}(y;x)\{1 - F_{Y|X}(y;x)\} \end{aligned}$$

its covariances are $Cov_{Y|X}(y,y';x) = \kappa_2 f_X^{-1}(x)\{F_{Y|X}(y \wedge y';x) - F_{Y|X}(y;x)F_{Y|X}(y';x)\}$ *and zero for* $x \neq x'$.

The weak convergence of Proposition 5.2 is still satisfied with the convergence rate $(Th_T)^{1/2}$ and the notations of Proposition 5.6. The quantile processes of Section 5.2 are generalized to the continuous process $(X_t, Y_t)_{t>0}$ and their asymptotic behaviour is deduced by the same arguments from the weak convergence of $(Th_T)^{1/2}(\widehat{F}_{Y|X,T,h_T} - F_{Y|X})$.

The conditional density $f_{Y|X}(y;x)$ of the ergodic limit of the process is estimated using the kernel K_{h_T}, with the same bandwidth as the estimator of the distribution function $F_{Y|X}$

$$\widehat{f}_{Y|X,T,h_T}(y;x) = \frac{1}{T}\int_0^T K_{h_T}(Y_t - y)\left\{\frac{\int_0^T K_{h_T}(x - X_s)1_{\{Y_s \le Y_t\}}}{\int_0^T K_{h_T}(x - X_s)\,ds}\right\} dt.$$

Its expectation is approximated by

$$\begin{aligned} f_{Y|X,T,h_T}(y;x) &= T^{-1}E\int_0^T K_{h_T}(Y_t - y)\{F_{Y|X}(Y_t;x) + h_T^2 b_F(Y_t;x)\}\,dt \\ &\qquad + o(h_T^2) \\ &= E\int_{\mathcal{I}_Y} K_{h_T}(v - y)\frac{\partial}{\partial v}\{F_{Y|X} + h_T^2 b_F\}(v;x)\,dv + o(h_T^2) \\ &= f_{Y|X}(y;x) + h_T^2\frac{\partial b_F(y;x)}{\partial y} + \frac{h_T^2}{2} m_{2K}\frac{\partial^2 f_{Y|X}(y;x)}{\partial y^2} \\ &\qquad + o(h_T^2), \end{aligned}$$

where b_F is the bias (5.6) of the estimator $\widehat{F}_{Y|X,n,h}$. Let v_f be defined by (5.17), the variance of $\widehat{f}_{Y|X,T,h_T}(y;x)$ has an expansion similar to the variance of the estimator $\widehat{f}_{Y|X,n,h}(y;x)$

$$v_{f_{Y|X},T,h_T} = (Th_T^2)^{-1}v_f(y;x) + o((Th_T^2)^{-1}).$$

Proposition 5.7. *Under the ergodicity conditions and for a conditional density* $f_{Y|X}$ *in class* $C_s(\mathcal{I}_{XY})$, *the bias and the variance of the estimator*

$\widehat{f}_{Y|X,T,h_T}$ are

$$b_{f_{Y|X},T,h_T}(y;x) = h_T^s b_{f_{Y|X}}(y;x) + o(h_T^s),$$
$$b_{f_{Y|X}}(y;x) = \frac{m_{sK}}{s!} f_X^{-1}(x)\{\partial^s f(x,y)/\partial x^s - f_{Y|X} f_X^{(s)}(x)\},$$
$$v_{f_{Y|X},T,h_T}(y;x) = (Th_T^2)^{-1}\{v_f(y;x) + o(1)\},$$
$$v_f(y;x) = \kappa_2 f_X^{-1}(x) f_{Y|X}(y;x)\{1 - f_{Y|X}(y;x)\}$$

its covariances are zero for $x \neq x'$ or $y \neq y'$.

The process $T^{1/2}h_T(\widehat{f}_{Y|X,T,h_T} - f_{Y|X})$ converges weakly to a Gaussian process with mean $\lim_T T^{1/2}h_T^{s+1} b_{f_{Y|X}}$, variance v_f and covariances are zero.

The optimal bandwidth for $\widehat{f}_{Y|X,T,h_T}$ is $O(T^{-1/(2s+2)})$ and the convergence rate of $\widehat{f}_{Y|X,T,h_T}$ with the optimal bandwidth is $T^{s/(2s+2)}$, hence it is $T^{1/3}$ for $s = 2$, and the expression of the optimal bandwidth is $h_{T,opt,f_{Y|X}}$ defined in the previous section.

5.6 Inverse of a regression function

Consider the inverse function (5.4) for a regression function m of the model (1.6), monotone on a sub-interval $\mathcal{I}_m$ of the support $\mathcal{I}_X$ of the regression variable X. The kernel estimator of the function m is monotone on the same interval with a probability converging to 1 as n tends to infinity, by an extension of Lemma 5.1 to an increasing function. The maxima and minima of the estimated regression function, considered in Section 3.7, define empirical intervals for monotonicity where the inverse of the regression function is estimated by the inverse of its estimator. Let t belong to the image $\mathcal{J}_m$ by m of an interval $\mathcal{I}_m$ where m is increasing

$$\widehat{Q}_{m,n,h}(t) = \widehat{m}_{n,h}^{-1}(t) = \inf\{x \in \mathcal{I}_m : \widehat{m}_{n,h}(x) \geq t\}. \tag{5.19}$$

This estimator is continuous like $\widehat{m}_{n,h}$, so that $\widehat{m}_{n,h} \circ \widehat{m}_{n,h}^{-1} = id$ on $\mathcal{J}_m$, and $\widehat{m}_{n,h}^{-1} \circ \widehat{m}_{n,h} = id$ on $\mathcal{I}_m$. The results proved in Section 5.2 for the conditional quantiles adapt to the estimator $\widehat{Q}_{m,n,h}$. The bias and the variance of the estimator (5.19) on $\mathcal{J}_m$ are deduced from those of the estimator $\widehat{m}_{n,h}$, as in Proposition 5.3

$$b_{Q_m,n,h}(t) = -h^2 \frac{b_m}{m^{(1)}} \circ Q_m(t) + o(h^2),$$

$$v_{Q_m,n,h}(t) = (nh)^{-1} \frac{\sigma_m^2}{m^{(1)2}} \circ Q_m(t) + o((nh)^{-1}).$$

The weak convergence of $(nh)^{1/2}(\widehat{Q}_{m,n,h} - Q_m)$ is a consequence of Theorem 3.1 and it is proved by the same arguments as Theorem 5.1 and proved by the same arguments. Let W_1 be the Gaussian process limit of $\sigma_m^{-1}(nh)^{1/2}(\widehat{m}_{n,h} - m_{n,h})$ on $\mathcal{I}_m$.

Theorem 5.3. *On $\mathcal{J}_m$, the process $U_{Q_m,n,h} = (nh)^{1/2}\{\widehat{Q}_{m,n,h} - Q_m\}$ converges weakly to $U_{Q_m} = \dfrac{W_1 + \gamma^{1/2} b_m}{m^{(1)}} \circ Q_m$.*

The inverse of the estimator (3.22) for a regression function of an ergodic and mixing process $(X_t, Y_t)_{t\geq 0}$ is $(Th_T)^{1/2}$-consistent and it satisfies the same approximations and weak convergence, with the notations and conditions of Section 3.10.

Under derivability conditions for the kernel, the regression function and the density of the variable X, the estimators $\widehat{m}_{n,h}$ and its inverse are differentiable and they belong to the same class which is supposed to be sufficiently large to allow expansions of order s for estimator of function m in C_{k+s}. The derivatives of the quantile are determined by consecutive derivatives of the inverse: $\widehat{Q}^{(1)}_{m,n,h} = \{\widehat{m}^{(1)}_{n,h} \circ \widehat{Q}_{m,n,h}\}^{-1}$, $\widehat{Q}^{(2)}_{m,n,h} = \{\widehat{m}^{(2)}_{n,h}\{\widehat{m}^{(1)3}_{n,h}\}^{-1}\} \circ \widehat{Q}_{m,n,h}$.

Consider a partition of the sample in J disjoint groups of size n_j, and let A_j be the indicator of a group j, for $j = 1, \ldots, J$. Let $Y = \sum_{j=1}^J Y_j 1_{A_j}$ and $X = \sum_{j=1}^J X_j 1_{A_j}$ where (X_j, Y_j) is the variable set in group j. For $j = 1, \ldots, J$, the regression model for the variables $(X_{ji}, Y_{ji})_{i=1,\ldots,n_j}$ is

$$Y_{ji} = m_j(X_{ji}) + \varepsilon_{ji}$$

where $m_j(x) = E(Y \mid 1_{A_j} X = x)$ and the expectation in the whole sample is defined from the probability p_j of A_j, the conditional density of X given A_j and the conditional regression functions given the group A_j

$$m(x) = \sum_{j=1}^J \pi_j(x) m_j(x),$$

$$\pi_j(x) = p_j \frac{f_j(x)}{f(x)} = P(A_j \mid X = x).$$

The density of X in the whole sample is a mixture of J densities conditionally on the group $f_X(x) = \sum_{j=1}^J p_j f_j(x)$ and the ratio $f^{-1}(x) f_j(x)$ is one if the partition is independent of X. The regression functions and the conditional probability densities are estimated from the sub-samples ,

$$\widehat{m}_{j,n,h}(x) = \frac{\sum_{i=1}^{n_j} Y_{ji} K_h(x - X_{ji})}{\sum_{i=1}^{n_j} K_h(x - X_{ji})},$$

$$\widehat{m}_{n,h}(x) = \frac{\sum_{j=1}^{J} \sum_{i=1}^{n_j} Y_{ji} K_h(x - X_{ji})}{\sum_{j=1}^{J} \sum_{i=1}^{n_j} K_h(x - X_{ji})},$$

$$\widehat{\pi}_{j,n,h}(x) = \frac{\sum_{i=1}^{n_j} K_h(x - X_{ji})}{\sum_{j=1}^{J} \sum_{i=1}^{n_j} K_h(x - X_{ji})}.$$

The inverse processes $\widehat{Q}_{j,m,n,h}$ are defined as in Equation (5.19) for each group. The inverse of the conditional probability densities π_j are estimated using the same arguments.

5.7 Quantile function of right-censored variables

The product-limit estimator $\widehat{F}_n$ for a differentiable distribution function F on $\mathbb{R}_+$ under right-censorship satisfies Equation (2.11) on $[0, \max T_i[$

$$\widehat{F}_n(x) = F - \{1 - F(x)\} \int_0^x \frac{1 - \widehat{F}_n^R(s^-)}{1 - F(s)} \, d(\widehat{\Lambda}_n - \Lambda)(s),$$

denoted $F - \psi_n$ where $E\psi_n = 0$ and $\sup_{t \le \tau} \|\psi_n(t)\|_2$ converges *a.s.* to zero for every $\tau < \tau_F = \sup\{x > 0; F(x) < 1\}$. Its quantile $\widehat{Q}_n$ converges therefore to Q_F in probability uniformly on $[0, 1[$. Let f be the density probability for F and let G be the distribution function of the independent censoring times, the process $n^{1/2}\psi_n$ converges weakly to a centered Gaussian process with covariance function

$$C_F(x, y) = \{1 - F(x)\}\{1 - F(y)\} \int_0^{x \wedge y} \{(1 - F)(1 - G^-)\}^{-1} \, dF,$$

at every x and y in $[0, \tau < \tau_0]$, where $\tau_0 = \tau_F \wedge \tau_G$. As a consequence, the quantile process

$$n^{1/2}(\widehat{Q}_n - Q_F) = -n^{1/2} \frac{\psi_n}{f} \circ Q_F + r_n \tag{5.20}$$

is unbiased and it converges weakly to a centered Gaussian process with covariance function $c(s, t) = C_F(Q_F(s), Q_F(t))\{f \circ Q_F(s)\}^{-1}\{f \circ Q_F(t)\}^{-1}$, for every s and t in $[0, F(\tau_0)]$. The remainder term is such that $\sup_{t \le F(\tau_0)} \|r_n\|$ is a $o_{L_2}(1)$.

A smoothed quantile process is defined by integrating the smoothed process $\int K_h(t-s)\,d\widehat{Q}_n(s)$ which is an uniformly consistent estimator of the derivative $Q_F^{(1)}(t) = 1/\{f \circ Q_F(t)\}$ of $Q_F(t)$. Its mean is $\int K_h(t-s)\,dQ(s)$ and its bias $b_{Q_F,n,h} = \frac{h^2}{2} m_{2K} Q_F^{(3)}(t)$ if F belongs to C_3. Its variance and covariance functions are deduced from the representation (5.20) of the quantile, for $s \neq t$ and as n tends to infinity

$$\begin{aligned}
Cov\{\widehat{Q}_{n,h}(t), \widehat{Q}_{n,h}(s)\} &= n^{-1} \int_0^t \int_0^s \int_{-1}^1 \int_{-1}^1 1_{\{u \neq v\}} K_h(u-u') \\
&\quad \times K_h(v-v')\,du\,dv\,d^2c(u',v') + (nh)^{-1}\kappa_2 c(s \wedge t, s \wedge t) + o(n^{-1/2}) \\
&= (nh)^{-1}\kappa_2 c(s \wedge t, s \wedge t) + o(n^{-1/2}).
\end{aligned}$$

5.8 Conditional quantiles with variable bandwidth

The pointwise conditional mean squared errors for the empirical conditional distribution function and its inverses reach their minimum at a varying bandwidth function. So the behaviour of the estimators with such bandwidth is now considered. Conditions 4.1 are supposed to be satisfied in addition to 2.1 or 2.2. The results of Propositions 5.1 and 5.2 still hold with a functional bandwidth sequence h_n and approximation orders $o(\|h_n\|^2)$ for the bias and $o(n\|h_n^{-1}\|)$ for the variance and a functional convergence rate $(nh_n)^{1/2}$ for the process $\nu_{Y|X,n,h}$. This is an application of Section 4.3 with the following expansion of the covariances.

Lemma 5.2. *The covariance of $\widehat{F}_{Y|X,n,h}(y;x_1)$ and $\widehat{F}_{Y|X,n,h}(y;x_2)$ equals*

$$\begin{aligned}
&2[n\kappa_2\{h_n(x_1) + h_n(x_2)\}]^{-1} \\
&\qquad \times \left[v_F(y; z_n(x_1,x_2)) \int K(v - \alpha_n(v))K(v + \alpha_n(v))\,dv \right. \\
&\qquad + \delta_n(x_1,x_2) f_X^{-2}(z_n(x,y))\{(v_\wedge Y|X f_X)^{(1)} - m^2 f_X^{(1)}\}(z_n(x,y)) \\
&\qquad \left. \times \int vK(v-\alpha_n(v))K(v+\alpha_n(v))\,dv + o(\|h_n\|)\right].
\end{aligned}$$

The mean squared errors the functional bandwidth sequences are similar to the MSE and MISE of Section 5.3.

The conditional quantiles are now defined with functional bandwidths satisfying the convergence condition 4.1. The representation of the conditional quantiles becomes

$$\widehat{Q}_{Y,n,h} = Q_Y + (\widehat{\eta}_{Y,n,h} + F_{Y|X} \circ \widetilde{Q}_{Y,n,h} - F_{Y|X} \circ Q_Y)(f_{Y|X} \circ Q_Y)^{-1} + o_{L_2}((n\|h_n\|)^{-1/2}),$$

$$\widehat{Q}_{X,n,h} = Q_X + (\widehat{\zeta}_{X,n,h} + F_{Y|X} \circ \widetilde{Q}_{X,n,h} - F_{Y|X} \circ Q_X)(f_{Y|X} \circ Q_X)^{-1} + o_{L_2}((n\|h_n\|)^{-1/2})$$

with $\widehat{\eta}_{Y,n,h}$ defined by (5.8) and $\widehat{\zeta}_{X,n,h} = F_{Y|X} \circ \widehat{Q}_{X,n,h} - F_{Y|X} \circ \widetilde{Q}_{X,n,h}$. The expansions of their bias and variance are also written with the uniform norms of the bandwidths, generalizing Propositions 5.3 and 5.4 and the weak convergence of the quantile processes is proved as for the kernel regression function with variable bandwidth in Section 4.4.

5.9 Exercises

(1) Consider the quantile process $\widehat{F}_n^{-1}$ of a continuous distribution function F and the smooth quantile estimator $\widehat{T}_n(t) = \int_0^1 K_h(t-s)\widehat{F}_n^{-1}(s)\,ds$, for t in $[0,1]$. Prove its consistency and write expansions for its bias and its variance.

(2) Determine the limiting distribution of the quantiles with respect to X and Y for the estimator of the distribution function of $Y \leq y$ conditionally on $X \leq x$.

(3) Determine the limiting distribution of smoothed quantiles with respect to X and Y for the estimator of the distribution function of $Y \leq y$ conditionally on $X \leq x$.

Chapter 6

Nonparametric estimation of intensities for stochastic processes

6.1 Introduction

Let $N_n = \{N_n(t), t \geq 0\}$ be a sequence of counting processes defined on a probability space $(\Omega, \mathcal{A}, P)$ associated to a sequence of random time variables $(T_i)_{1 \leq i \leq n}$

$$N_n(t) = \sum_{i=1}^{n} 1_{\{T_i \leq t\}}, t \geq 0,$$

where $T_i = \inf\{t; N_n(t) = i\}$, and let $\mathcal{F}_n = (\mathcal{F}_{nt})_{t \in \mathbb{R}_+}$ denote the history generated by observations of N_n and other observed processes before t. The predictable compensator of N_n with respect to $\mathcal{F}_n$ is the unique $\mathcal{F}_n^-$-measurable (or *predictable*) process $\widetilde{N}_n$ such that $N_n - \widetilde{N}_n$ is a $\mathcal{F}_n$-martingale on $(\Omega, \mathcal{A}, P)$. Consider a counting process N_n with a predictable compensator

$$\widetilde{N}_n(t) = \sum_{i=1}^{n} \int_0^t Y_i(s) \mu(s, Z_i(s))\, ds$$

where Y_i and Z_i are predictable processes with values in metric spaces $\mathcal{Y}$ and $\mathcal{Z}$ and $\mu(s, z) = \lambda(s) r(z)$ is a strictly positive function for $s > 0$. This model with a random variable or process Z is classical, when the observations are a right-censored counting process N_n and $Y_n(t) = \sum_{i=1}^{n} 1_{\{T_i \geq t\}}$ is the counting process of the random times of N_n after t. The right-censorship is defined by a sequence of random censoring variables $(C_i)_{1 \leq i \leq n}$ so that a censoring process $N_n^C(t) = \sum_{i=1}^{n} 1_{\{C_i \leq t\}}$ is partially observed with the processes N_n, the observations are the processes $Y_n(t) = \sum_{1 \leq i \leq n} 1_{\{T_i \wedge C_i \geq t\}}$ and the sequences of times $(T_i \wedge C_i)_{1 \leq i \leq n}$ and indicators $(\delta_i)_i = (1_{\{T_i \wedge C_i\}})_i$ with values 1 if T_i is observed and 0 otherwise. All processes are observed in an increasing time interval $[0, \tau]$ such that $N_n(\tau) = n$ tends to infinity.

With independent and identically distributed variables T_i with distribution function F_T and density f_T, the relationships between the survival function $1 - F_T(t) = \exp\{-\int_0^t \lambda(s)\,ds\}$ and the hazard function $\lambda_T = (1 - F_T)^{-1} f_T$ are equivalent. With independent and identically distributed censoring variables C_i, independent of the time sequence $(T_i)_{1\le i\le n}$ and with distribution function F_C, the hazard function of the censored counting process $N_n(t) = \sum_{1\le i\le n} \delta_i 1_{\{T_i\le t\}}$ is identical to λ_T. The aim of this chapter is to define smooth estimators for the baseline hazard function and regression function of intensity models and to compare them with histogram-type estimators. Several regression models are considered, with parametric or nonparametric regression functions.

Let $J_n(t) = 1_{\{Y_n(t)>0\}}$ be the indicator of censored times occurring after t. The baseline intensity λ of an intensity $\mu_n(t) = \lambda(t)Y_n(t)$ is estimated for t in $[h, \tau - h]$ by smoothing the Nelson (1972) estimator of the cumulative hazard function $\Lambda(t) = \int_0^t \lambda(s)\,ds$, which is asymptotically equivalent to $\int_0^t J_n(s)Y_n^{-1}(s)\,d\widetilde{N}_n(s)$ as J_n tends to 1 in probability. The unbiased Nelson estimator is defined as $\widehat{\Lambda}_n(t) = \int_0^t J_n(s)Y_n^{-1}(s)\,dN_n(s)$, with the convention $0/0 = 0$, and the function λ is estimated by smoothing $\widehat{\Lambda}_n$

$$\widehat{\lambda}_{n,h}(t) = \int_{-1}^{1} Y_n^{-1}(s)J_n(s)K_h(t-s)\,dN_n(s). \tag{6.1}$$

A stepwise estimator for λ is also defined on an observation time $[0, \tau]$ as the ratio of integrals over the subintervals $(B_{jh})_{j\le J_h}$ of a partition of the observation interval into $J_h = h^{-1}\tau$ disjoint intervals with length h tending to zero. For every t belonging to B_{jh}, the histogram-type estimator of the funtion λ is estimated at t by

$$\widetilde{\lambda}_{n,h}(t) = \{\int_{B_{jh}} J_n(s)\,dN_n\}\{\int_{B_{jh}} Y_n(s)\,ds\}^{-1} \tag{6.2}$$

where the normalizing h of the histogram for a density is replaced by an integral. Consider a multiplicative intensity $\mu(t, Z_i(t)) = \lambda(t)r(Z_i(t))Y_i(t)$ for the counting process $N_i(t) = 1_{\{T_i\le t\}}$ of the i-th time variable. In the Cox model, the regression function r defining the point process is $\exp(\beta^T z)$, with an unknown parameter β belonging to an open bounded set of $\mathbb{R}^d$ and z in the metric space $(\mathcal{Z}, \|\cdot,\|)$ of the sample-paths of a regression processes Z_i. The intensity conditionally on $\mathcal{F}_t$ is then the semi-parametric process

$$\lambda(t;\beta) = \sum_{1\le i\le n} 1_{\{T_i\le t\}} r(Z_i(t))\lambda(t)$$

and the estimators of λ and β are defined by the means of the process $S_n^{(0)}$ defined by weighting each term of the sum in Y_T by the regression function at the jump time

$$S_n^{(0)}(t;\beta) = \sum_{i=1}^{n} r_{Z_i}(t;\beta) 1_{\{T_i \geq t\}},$$

with the parametric function $r_Z(t;\beta) = \exp\{\beta^T Z(t)\}$. For $k = 1, 2$, let also $S_n^{(0)}(t;\beta) = \sum_{i=1}^{n} r_{Z_i}(t;\beta) Z_i^{\otimes k}(t) 1_{\{T_i \geq t\}}$ be the derivatives of $S_n^{(0)}$ with respect to β, let $Z^{\otimes 0} = 1$, $Z^{\otimes 1} = Z$ and $Z^{\otimes 2}$ be the scalar product. The true regression parameter value is β_0, or r_0 for the function r and the predictable compensator of the point process N_n is

$$\widetilde{N}_n(t) = \int_0^t S_n^{(0)}(s;\beta)\lambda(s)\, ds\,. \tag{6.3}$$

The function λ is estimated by smoothing the cumulative hazard function of the Cox process, the parameter β by maximizing the partial likelihood

$$\widehat{\lambda}_{n,h}(t;\beta) = \int_{-1}^{1} J_n(s)\{S_n^{(0)}(s;\beta)\}^{-1} K_h(t-s)\, dN_n(s), \tag{6.4}$$

$$\widehat{\beta}_{n,h} = \arg\max_{\beta} \prod_{T_i \leq \tau} \{r_{Z_i}(t;\beta)\widehat{\lambda}_{n,h}(T_i;\beta)\}^{\delta_i},$$

with the convention $0^0 = 1$ and $J_n = 1_{\{Y_n > 0\}}$. The hazard function is estimated by $\widehat{\lambda}_{n,h} = \widehat{\lambda}_{n,h}(\widehat{\beta}_{n,h})$. The classical estimators of the Cox model rely on the estimation of the function $\Lambda(t)$ by the stepwise process $\widehat{\Lambda}_n(t;\beta) = \int_0^\tau J_n(s)\{S_n^{(0)}(s;\beta)\}^{-1} dN_n(s)$ at fixed β and the parameter β of the exponential regression function $r_Z(t;\beta) = e^{\beta^T Z(t)}$ is estimated by maximization of an expression similar to (6.4) where $\widehat{\lambda}_{n,h}(T_i;\beta)$ is replaced by the jump of $\widehat{\Lambda}_n(\beta)$ at T_i

$$\widehat{\beta}_n^C = \arg\max_{\beta} \prod_{T_i \leq \tau} \{r_{Z_i}(t;\beta) S_n^{(0)-1}(T_i;\beta)\}^{\delta_i}.$$

A stepwise estimator for the baseline intensity is now defined as

$$\widetilde{\lambda}_{n,h}(t;\beta) = \sum_{j \leq J_h} 1_{B_{jh}}(t)\Big[\int_{B_{jh}} J_n(s)\, dN_n(s)\Big]\Big[\int_{B_{jh}} S_n^{(0)}(s;\beta)\, ds\Big]^{-1}, \tag{6.5}$$

$$\widetilde{\beta}_{n,h} = \arg\max_{\beta} \prod_{T_i \leq \tau} \{r_{Z_i}(T_i;\beta)\widetilde{\lambda}_{n,h}(T_i;\beta)\}^{\delta_i},$$

and $\widetilde{\lambda}_{n,h} = \widetilde{\lambda}_{n,h}(\widetilde{\beta}_{n,h})$. More generally, a nonparametric function r is estimated by a stepwise process $\widehat{r}_{n,h}$ defined on each set B_{jh} of the partition

$(B_{jh})_{j\le J_h}$, centered at a_{jh}. Let also $(D_{lh})_{l\le L_h}$ be a partition of the values $Z_i(t)$, $i \le n$, centered at z_{lh}. The function r is estimated in the form $\widetilde{r}_{n,h}(Z(t)) = \sum_{l\le L_h} \widetilde{r}_{n,h}(z_{lh}) 1_{D_{lh}}(Z(t))$

$$\widetilde{\lambda}_{n,h}(t;r) = \sum_{j\le J_h} 1_{B_{jh}}(t) \int_{B_{jh}} J_n(s)\,dN_n(s)\,[\int_{B_{jh}} S_n^{(0)}(s;r)\,ds]^{-1},$$

$$\widetilde{r}_{n,h}(z_{lh}) = \arg\max_{r_l} \prod_{T_i\le\tau} [\{\sum_{l\le L_h} r_l 1_{D_{lh}}(Z_i(T_i))\}\widetilde{\lambda}_{n,h}(T_i;r_{lh})]^{\delta_i}, \quad (6.6)$$

where $S_n^{(0)}(t;r) = \sum_{i=1}^n r_{Z_i}(t) 1_{\{T_i\ge t\}}$ is now defined for a nonparametric regression function, then $\widetilde{\lambda}_{n,h}(t,Z) = \widetilde{\lambda}_{n,h}(t,\widetilde{r}_n(Z(t)))$. A kernel estimator for the functions λ is similarly defined by

$$\widehat{\lambda}_{n,h}(t;r) = \int_{-1}^{1} J_n(s)\{S_n^{(0)}(s;r)\}^{-1} K_h(t-s)\,dN_n(s)\,. \quad (6.7)$$

An approximation of the covariates values at jump times by z when they are sufficiently close allows to build a nonparametric estimator of the regression function r like β in the parametric model for r

$$\widehat{r}_{n,h}(z) = \arg\max_{r_z} \sum_{i=1}^n \int_{-1}^{1} \{\log r_z(s) + \log \widehat{\lambda}_n(s;r_z)\} K_{h_2}(z - Z_i(s))\,dN_i(s),$$

where $h_2 = h_{n2}$ is a bandwidth sequence satisfying the same conditions as h, and $\widehat{\lambda}_n(t,Z) = \widehat{\lambda}_n(t;\widehat{r}_n(t,Z(t))$.

The L_2-risk of the estimators of the intensity functions splits into a squared bias and a variance term and the minimization of the quadratic risk provides an optimal bandwidth depending on the parameters and functions of the models and having similar rates of convergence, following the same arguments as for the optimal bandwidths for densities.

6.2 Risks and convergences for estimators of the intensity

Conditions for expanding the bias and variance of the estimators are added to Conditions 2.1 and 2.2 of the previous chapters concerning the kernel and the bandwidths. For the intensities, they are regularity and boundedness conditions for the functions of the models and for the processes.

Condition 6.1.

(1) As n tends to infinity, the process Y_n is positive with a probability tending to 1, *i.e.* $P\{\inf_{[0,\tau]} Y_n > 0\}$ tends to 1, and there exists a function g such that $\sup_{[0,\tau]} |n^{-1}Y_n - g|$ tends a.s. to zero;

(2) $\int_0^\tau g^{-1}(s)\lambda(s)\,ds < \infty$
(3) The functions λ and g belong to $C_s(\mathbb{R}_+)$, $s \geq 2$.

6.2.1 *Kernel estimator of the intensity*

Let

$$\lambda_{n,h}(t) = \int_{-1}^{1} J_n(s) K_h(t-s)\lambda(s)\,ds, \quad \text{for } t \in [h, \tau - h],$$

be the expectation of the kernel estimator (6.1) and let $\lambda_n(t) = J_n(t)\lambda(t)$, defined as $\lambda(t)$ on the random interval $\mathcal{I}_{n,\tau} = 1_{\{J_n=1\}}[0,\tau]$ which may be right-censored. Let also $\mathcal{I}_{n,h,\tau} = 1_{\{J_n=1\}}[h, \tau-h]$ be the interval where all convergences will be considered. Since $J_n(t) - 1$ tends uniformly to zero in probability, $\sup_{t\in[0,\tau]} |\lambda(t) - \lambda_{n,h}(t)|$ tends to zero in probability.

Proposition 6.1. *Under Conditions 2.1 and 6.1 with h_n converging to zero and nh to infinity*
(a) $\sup_{t\in\mathcal{I}_{n,h,\tau}} |\widehat{\lambda}_{n,h}(t) - \lambda(t)|$ converges to zero in probability. (b) For every t in $\mathcal{I}_{n,h,\tau}$, the bias $\lambda_{n,h}(t) - \lambda(t)$ of the estimator is

$$b_{\lambda,n,h}(t;s) = \frac{h^s}{s!} m_{sK} \lambda^{(s)}(t) + o(h^s),$$

denoted $h^2 b_\lambda(t) + o(h^2)$, its variance is

$$Var\{\widehat{\lambda}_{n,h}(t)\} = (nh)^{-1}\kappa_2\, g^{-1}(t)\lambda(t) + o((nh)^{-1}),$$

also denoted $(nh)^{-1}\sigma_\lambda^2(t) + o((nh)^{-1})$ and the covariance of $\widehat{\lambda}_{n,h}(t)$ and $\widehat{\lambda}_{n,h}(s)$ is

$$Cov\{\widehat{\lambda}_{n,h}(s), \widehat{\lambda}_{n,h}(t)\} = (nh)^{-1}\{\frac{\lambda}{g}(\frac{s+t}{2}) \int K(v-\alpha_h)K(v+\alpha_h)dv + o(1)\}$$

if $\alpha_h = |t-s|/2h \leq 1$ and zero otherwise, with uniform approximations on $\mathcal{I}_{n,h,\tau}$. The L_p-norms of the estimator are

$$\sup_{t\in\mathcal{I}_{n,h,\tau}} \|\widehat{\lambda}_{n,h}(t) - \lambda_{n,h}(t)\|_p = 0((nh)^{-1/p})$$

and $\sup_{t\in\mathcal{I}_{n,h,\tau}} \|\widehat{\lambda}_{n,h}(t) - \lambda(t)\|_p = 0((nh)^{-1/p} + h^s)$.

Proof. (a) Let $M_{\Lambda,n} = \widehat{\Lambda}_n - \Lambda$, its predictable compensator is $\widetilde{N}_n$. By Lenglart's inequality applied to the martingale $M_{\Lambda,n}$, for every $n \geq 1$

$$P(\sup_{[h,\tau-h]} |\widehat{\lambda}_{n,h} - \lambda| \geq \eta) \leq \eta^{-2} E \sup_{t\in[h,\tau-h]} \int_{-1}^{1} K_h^2(t-u)(J_n Y_n^{-1}\lambda)(u)\,du$$
$$= \eta^{-2}(nh)^{-1}\kappa_2 \sup_{t\in[0,\tau]} g^{-1}(t)\lambda(t)$$

and the upper bound tends to zero as n tends to infinity.

(b) For every t in $\mathcal{I}_{n,h,\tau}$, the bias $b_{\lambda,n,h}(t) = \lambda_h(t) - \lambda_{n,h}(t)$ develops as

$$\begin{aligned} b_{\lambda,n,h}(t) &= \int_{-1}^{1} K_h(t-u) E\{J_n(u)\lambda(u) - J_n(t)\lambda(t)\}\, du \\ &= \int_{-1}^{1} E\{J_n(t+hz)\lambda(t+hz) - J_n(t)\lambda(t)\} K(z)\, dz \\ &= \frac{h^s}{s!}\lambda^{(s)}(t) + o(h^s), \end{aligned}$$

where $EY_n(s) = P(Y_n(s) > 0) = P(\max_{i\le n} T_i > s) = 1 - F_T^n(s)$ belongs to $]0,1[$ for every $s \le \tau$, for independent times T_i, $i \le n$. Its variance is

$$\begin{aligned} Var\{\widehat{\lambda}_{n,h}(t)\} &= E\int_{-1}^{1} K_h^2(t-u) J_n(u) Y_n^{-1}(u)\lambda(u)\, du \\ &= (nh)^{-1} \int K^2(z) E J_n(t+hz) g^{-1}(t+hz)\lambda(t+hz)\, dz \\ &= (nh)^{-1}\kappa_2 g^{-1}(t)\lambda(t) + o((nh)^{-1}). \end{aligned}$$

The covariance of $\widehat{\lambda}_{n,h}(t)$ and $\widehat{\lambda}_{n,h}(s)$ is $\int K_h(s-u)K_h(t-u)J_n(u)Y_n^{-1}(u)\lambda(u)\, du$, it is zero if $\alpha_h = |x-y|/(2h) > 1$ and, if $\alpha_h \le 1$, it is approximated by a change of variables as in Proposition 2.2 for the density, under Conditions 6.1.

The L_p-moment of $\widehat{\lambda}_{n,h}(t) - \lambda_{n,h}(t) = \int_{-1}^{1} K_h(t-u) J_n(u) Y_n^{-1}(u)\, d(N_n - \widetilde{N}_n)(u)$ are calculated using the martingale property of $M_n = N_n - \widetilde{N}_n$ and its stochastic integrals

$$\begin{aligned} E|\widehat{\lambda}_{n,h}(t) - \lambda_{n,h}(t)|^3 &= (nh)^{-1}\{\kappa_2 g^{-1}(t)\lambda^2(t) + o(1)\}, \\ E|\widehat{\lambda}_{n,h}(t) - \lambda_{n,h}(t)|^4 &= (nh)^{-1}\{\kappa_2 g^{-1}(t)\lambda^3(t) + o(1)\}, \\ E|\widehat{\lambda}_{n,h}(t) - \lambda_{n,h}(t)|^5 &= (nh)^{-1}\{\kappa_2 g^{-1}(t)\lambda^4(t) + o(1)\}. \end{aligned}$$

The higher order moments are deduced by iterations. In each case, the main term is expressed as the integral of the product of a squared kernel at a time T_i and other kernel terms at T_{j_k}, where all time variables are independent. □

For $p = 2$, $E\{\widehat{\lambda}_{n,h}(t) - \lambda_{n,h}(t)\}^2$ develops on $\mathcal{I}_{n,h,\tau}$ as the sum of a squared bias and a variance terms $\{\lambda_{n,h}(t) - \lambda(t)\}^2 + Var\{\widehat{\lambda}_{n,h}(t)\}$ and its first order expansions are $(s!)^{-2} m_{sK}^2 h^{2s} \lambda^{(s)2}(t) + o(h^{2s}) + (nh)^{-1}\kappa_2 g^{-1}(t)\lambda(t) + o(n^{-1}h^{-1})$. The asymptotic mean squared error

$$AMSE(t;\widehat{\lambda}_{n,h}) = (nh)^{-1}\kappa_2 g^{-1}(t)\lambda(t) + \frac{1}{(s!)^2} m_{sK}^2 h^{2s}\lambda^{(s)2}(t)$$

is minimum for the bandwidth function

$$h_{AMSE,n}(t) = n^{-1/(2s+1)}\Big\{\frac{s!(s-1)!\,\kappa_2\lambda(t)}{2m_{sK}^2 g(t)\lambda^{(s)2}(t)}\Big\}^{1/(2s+1)}.$$

The global asymptotic mean integrated squared error for $\widehat{\lambda}_{n,h}$ at t is

$$AMISE(\widehat{\lambda}_{n,h}) = (nh)^{-1}\kappa_2\int_0^{\mathcal{T}} g^{-1}(t)\lambda(t)\,dt + \frac{1}{(s!)^2}m_{sK}^2 h^{2s}\int_0^{\mathcal{T}} \lambda^{(s)2}(t)\,dt,$$

it is minimum for the global bandwidth

$$h_{AMISE,n} = n^{-1/(2s+1)}\Big\{\frac{s!(s-1)!\,\kappa_2\int_0^{\mathcal{T}} g^{-1}(t)\lambda(t)\,dt}{2m_{sK}^2\int_0^{\mathcal{T}}\lambda^{(s)2}(t)\,dt}\Big\}^{1/(2s+1)}.$$

It is estimated by

$$\widehat{h}_{AMISE,n,h} = n^{-1/(2s+1)}\Big[\frac{s!(s-1)!\,\kappa_2\int_0^{\mathcal{T}} Y_n^{-1}(t)\,d\widehat{\Lambda}_n(t)}{2m_{sK}^2\int_0^{\mathcal{T}}\{\widehat{\lambda}_{n,h}^{(s)}(t)\}^2\,dt}\Big]^{1/(2s+1)}.$$

Another integrated asymptotic mean squared error is the average of $AMSE(T)$ for the intensity, $E\{AMSE(T)\} = \int_0^{\mathcal{T}} AMSE\,dF_T$ also written

$$AMISE_n(h;F_T) = \int\{h^{2s}b_\lambda(t) + (nh)^{-1}\sigma_\lambda^2(t)\}\{1-F(t)\}\,d\Lambda(t)$$

and it is estimated by plugging estimators of the intensity, b_λ and σ_λ^2 into the empirical mean

$$n^{-1}\int_0^{\mathcal{T}}\{h^{2s}b_\lambda^2(t) + (nh)^{-1}\sigma_\lambda^2(t)\}\exp\{-\widehat{\Lambda}_{n,h}(t)\}Y_n^{-1}(t)\,dN_n(t).$$

Its minimum empirical bandwidth is

$$\widehat{h}_{\lambda,n,h} = n^{-1/(2s+1)}\Big\{\frac{\int_0^{\mathcal{T}}\widehat{\lambda}_{n,h}\exp\{-\widehat{\Lambda}_{n,h}\}Y_n^{-2}\,dN_n}{2m_{sK}^2\int_0^{\mathcal{T}}\{\widehat{\lambda}_{n,h}^{(s)}\}^2\exp\{-\widehat{\Lambda}_{n,h}\}Y_n^{-1}\,dN_n}\Big\}^{1/(2s+1)}.$$

An estimator of the derivative $\lambda^{(k)}$ or its integral are defined by the means of the derivatives $K^{(k)}$ of the kernel, for $k \geq 1$

$$\widehat{\lambda}_{n,h}^{(k)}(t) = \int K_h^{(k)}(t-s)J_n(s)Y_n^{-1}(s)\,dN_n(s).$$

Proposition 2.3 established for the densities is generalized to the intensity λ. Lemma 2.1 allows to develop the mean of the estimator of the first derivative as

$$\begin{aligned}\lambda_{n,h}^{(1)}(t) &= \int K_h^{(1)}(u-t)\lambda(u)\,du\\ &= -\lambda^{(1)}(t)\int zK^{(1)}(z)\,dz - \frac{h^2}{6}\lambda^{(3)}(t)\int z^3K^{(1)}(z)\,dz + o(h^2)\\ &= \lambda^{(1)}(t) + \frac{h^2}{2}m_{2K}\lambda^{(3)}(t) + o(h^2),\end{aligned}$$

for an intensity of C_3 or $\lambda^{(1)}_{n,h}(t) = \lambda^{(1)}(t) + \frac{h^s}{s!} m_{sK}\lambda^{(s)}(t) + o(h^s)$ for an intensity of C_s. The variance of $\widehat{\lambda}^{(1)}_{n,h}$ is $(nh^3)^{-1}g^{-1}(t)\lambda(t)\int K^{(1)2}(z)\,dz + o((nh^3)^{-1})$. The optimal local bandwidth for estimating $\lambda^{(1)}$ belonging to C_s is therefore

$$h_{AMSE}(\lambda^{(1)};t) = n^{-1/(2s+3)}\{s!(s-1)!\frac{\lambda(t)\int K^{(1)2}(z)\,dz}{2m_{sK}^2 g(t)\lambda^{(3)2}(t)}\}^{1/(2s+3)}.$$

For the second derivative of the intensity, the estimator $\widehat{\lambda}^{(2)}_{n,h}$ is the derivative of $\widehat{\lambda}^{(1)}_{n,h}$ expressed by the means of the second derivative of the kernel. For a function λ in class C_4, the expectation of the estimator $\widehat{\lambda}^{(2)}_{n,h}$ is $\lambda^{(2)}_{n,h}(t) = \lambda^{(2)}(t) + \frac{h^2}{2} m_{2K}\lambda^{(4)}(t) + o(h^2)$, so it converges uniformly to $\lambda^{(2)}$. The bias of $\widehat{\lambda}^{(2)}_{n,h}$ is $\frac{h^2}{2} m_{2K}\lambda^{(4)}(t) + o(h^2)$ and its variance $(nh^5)^{-1}g^{-1}(t)\lambda(t)\int K^{(2)2}(z)\,dz + o((nh^5)^{-1})$.

Proposition 6.2. *Under Conditions 2.2, for every integers $k \geq 0$ and $s \geq 2$ and for intensities belonging to class C_s, the estimator $\widehat{\lambda}^{(k)}_{n,h}$ of the k-order derivative $\lambda^{(k)}$ has a bias $O(h^s)$ and a variance $O((nh^{2k+1})^{-1})$ on $\mathcal{I}_{n,h,\tau}$. Its optimal local and global bandwidths are $O(n^{-s/(2k+2s+1)})$ and the optimal L_2-risks are $O(n^{-s/(2k+2s+1)})$.*

Consider the normalized process

$$U_{\lambda,n,h}(t) = (nh)^{1/2}\{\widehat{\lambda}_{n,h}(t) - \lambda(t)\},\ t \in \mathcal{I}_{n,h,\tau}.$$

The tightness and the weak convergence of $U_{\lambda,n,h}$ on $\mathcal{I}_{n,h,\tau}$ are proved by studing moments of its variations and the convergence of its finite dimensional distributions. For independent and identically distributed observations of right-censored variables, the intensity of the censored counting process has the same degree of derivability as the density functions for the random times of interest.

Lemma 6.1. *Under Conditions 6.1, for every intensity of C_s there exists a constant C such that for every t and t' in $\mathcal{I}_{n,h,\tau}$ satisfying $|t'-t| \leq 2h$*

$$Var\{\widehat{\lambda}_{n,h}(t) - \widehat{\lambda}_{n,h}(s)\}^2 \leq C(nh^3)^{-1}|t-t'|^2.$$

Proof. Let t' and t in $\mathcal{I}_{n,h,\tau}$, the variance of $\widehat{\lambda}_{n,h}(t') - \widehat{\lambda}_{n,h}(t)$ develops according to their variances given by Proposition 6.1 and the covariance between both terms which is zero if $|t-t'| > 1$ as established in the same proposition. The second order moment $E|\widehat{\lambda}_{n,h}(t) - \widehat{\lambda}_{n,h}(t')|^2$ develops as $\int\{K_h(t-u)-K_h(t'-u)\}^2 J_n(u)Y_n^{-1}(u)\lambda(u)\,du$ and it is a $O((t-t')^2 n^{-1}h_n^{-3})$, by the same approximation as for the proof of Lemma 2.2 and the uniform convergence of $J_nY_n^{-1}$. □

Theorem 6.1. *Under Conditions 6.1, for a density λ of class $C_s(\mathcal{I}_\tau)$ and with nh^{2s+1} converging to a constant γ, the process*

$$U_{\lambda,n,h} = (nh)^{1/2}\{\widehat{\lambda}_{n,h} - \lambda\}1_{\{\mathcal{I}_{n,h,\tau}\}}$$

converges weakly to $W_\lambda + \gamma^{1/2}b_\lambda$, where W_λ is a continuous Gaussian process on $\mathcal{I}_\tau$ with mean zero and covariance $E\{W_\lambda(t')W_\lambda(t)\} = \sigma^2_\lambda(t)\delta_{\{t',t\}}$, at t' and t in $\mathcal{I}_\tau$, and $\sigma^2_\lambda(t) = g^{-1}(t)\lambda(t)$.

Proof. The weak convergence of the finite dimensional distributions of the process $W_{\lambda,n,h}(t) = (nh)^{1/2}(\widehat{\lambda}_{n,h} - \lambda_{n,h})(t) = (nh)^{1/2}\int_{-1}^{1} K_h(t - u)J_n(u)Y_n^{-1}(u)\,dM_n(u)$ on $\mathcal{I}_{n,h,\tau}$ is a consequence of the convergence of its variance and of the weak convergence of the martingale $n^{-1/2}M_n$ to a continuous Gaussian process with mean zero and covariance $\int_0^{t\wedge t'} g(u)\lambda(u)\,du$ at t and t'. The covariance between $W_{\lambda,n,h}(t)$ and $W_{\lambda,n,h}(t')$, for $t \neq t'$, is approximated by

$$\begin{aligned} n^{-1} &\int K_h(t-u)K_h(t'-u)g(u)^{-1}\lambda(u)\,du \\ &= \frac{1_{\{0\le\alpha<1\}}}{2nh}\{\frac{\lambda}{g}(t) + \frac{\lambda}{g}(t')\}\Big(\int K(v-\alpha)K(v+\alpha)dv\Big) + o((nh)^{-1}), \end{aligned}$$

where $\alpha_h = |t' - t|/(2h)$ tends to infinity as h tends to zero, then the integral $\int K(v-\alpha_h)K(v+\alpha_h)dv$ tends to zero. Lemma 6.1 and the bound $E\{\lambda_{n,h}(t)-\lambda(t)-\lambda_{n,h}(t')+\lambda(t')\}^2 \le C'(nh)^{-1}|t-t'|^2$ imply that the mean of the squared variations of $U_{n,h}$ are $O(h^{-2}|t-t'|^2)$ if $|t-t'| \le 2h_n$ and $O(|t-t'|^2)$ if $|t-t'| > 2h_n$. The process $U_{n,h}$ is therefore tight. □

Corollary 6.1. *The process*

$$\sup_{t\in\mathcal{I}_{n,h,\tau}} \sigma_\lambda^{-1}(t)|U_{\lambda,n,h}(t) - \gamma^{1/2}b_\lambda(t)|$$

converges weakly to $\sup_{\mathcal{I}_\tau}|W_1|$, where W_1 is the Gaussian process with mean zero, variance 1 and covariances zero.
For every $\eta > 0$, there exists a constant $c_\eta > 0$ such that

$$\Pr\{\sup_{\mathcal{I}_{n,h,\tau}} |\sigma_\lambda^{-1}(U_{\lambda,n,h} - \gamma^{1/2}b_\lambda) - W_1| > c_\eta\}$$

tends to zero as n tends to infinity.

The Hellinger distance between two probability measures P_1 and P_2 defined by intensity functions λ_1 and λ_2 is

$$h^2(P_1,P_2) = \int \{1 - (\frac{\lambda_1}{\lambda_2})^{1/2}(\frac{1-F_1}{1-F_2})^{1/2}\}\,dF_1$$

and it is also written $h^2(P_1, P_2) = \int\{1 - (\frac{\lambda_1}{\lambda_2})^{1/2} e^{-(\Lambda_1-\Lambda_2)/2}\}\, dF_1$. The estimator of a function λ satisfies

$$\begin{aligned} h^2(\widehat{\lambda}_{n,h}, \lambda) &= \int \{1 - (\frac{\widehat{\lambda}_{n,h}}{\lambda} \frac{1-\widehat{F}_n}{1-F})^{1/2}\}\, dF \\ &\leq \int \{(\frac{\widehat{\lambda}_{n,h}}{\lambda} \frac{1-\widehat{F}_n}{1-F})^{1/2} - 1\}\, d(\widehat{F}_n - F) \end{aligned}$$

and the convergence rate of $h^2(\widehat{\lambda}_{n,h}, \lambda)$ to zero is $nh_n^{1/2}$.

A varying bandwidth estimator is defined for multiplicative intensities under Condition 4.1, with the optimal convergence rate. The bias and the variance of the estimator are modified as

$$b_{\lambda,n,h_n(t)}(t) = \frac{h_n(t)^s}{s!} m_{2K} \lambda^{(s)}(t) + o(\|h_n\|^2),$$

its variance is

$$Var\{\widehat{\lambda}_{n,h_n(t)}(t)\} = (nh_n(t))^{-1} \kappa_2\, g^{-1}(t)\lambda(t) + o((n^{-1}\|h_n^{-1}\|),$$

and $E\|\widehat{\lambda}_{n,h_n(t)}(t) - \lambda_{n,h_n(t)}(t)\|_p = 0((n^{-1}\|h_n^{-1}\|)^{1/p})$. The covariance of $\widehat{\lambda}_{n,h_n(t)}(t)$ and $\widehat{\lambda}_{n,h_n(t)}(t')\}$ equals

$$\begin{aligned} &E \int K_{h_n(t)}(t-u) K_{h_n(t')}(t'-u) Y_n^{-1}(u)\, d\Lambda(u) \\ &= \frac{2}{n\{h_n(t)+h_n(t')\}} \{(g^{-1}\lambda)(z_n(t,t')) \int K(v-\alpha_n(v))K(v+\alpha_n(v))\, dv \\ &+ \delta_n(x,y)(g^{-1}\lambda)^{(1)}(z_n(x,y)) \int vK(v-\alpha_n(v))K(v+\alpha_n(v))\, dv + o(\|h_n\|)\} \end{aligned}$$

with $\alpha_n(x,y,u) = \frac{1}{2}\{(u-x)h_n^{-1}(x) - (u-y)h_n^{-1}(y)\}$ and $v = \{(u-x)h_n^{-1}(x) + (u-y)h_n^{-1}(y)\}/2$. Lemma 4.2 is fulfilled for the mean squared variations of the process $\widehat{\lambda}_{n,h_n(t)}(t)$ which satisfy $E|\widehat{\lambda}_{n,h_n(t)}(t) - \widehat{\lambda}_{n,h(t')}(t')|^2 = O(n^{-1}\|h_n^{-1}\|^3(t-t')^2)$, if $|th_n^{-1}(t) - t'h_n^{-1}(t')| \leq 1$. Otherwise, the mean squared variations of $\widehat{\lambda}_{n,h}$ are zero, this implies the weak convergence of the process $(nh_n(t))^{1/2}\{\widehat{\lambda}_{n,h_n(t)}(t) - \lambda(t)\}I\{t \in \mathcal{I}_{n,\|h_n\|,\tau}\}$ to the process $W_f(t) + h^{1/2}(t)b_f(t)$, where W_f is a continuous centered Gaussian process on $\mathcal{I}_\tau$ with covariance $\sigma_\lambda^2(t)\delta_{\{t,t'\}}$ at t and t'.

6.2.2 *Histogram estimator of the intensity*

The histogram estimator (6.2) for the intensity λ is a consistent estimator as h tends to zero and n to infinity. Let $K_h(t) = h^{-1}\sum_{j \in J_{\tau,h}} 1_{B_{jh}}(t)$ be the kernel corresponding to the histogram, the histogram estimator (6.2) is defined as the ratio of two stochastic integrals on the same subintervals of the partition of $[0, \tau]$. Its expectation is approximated by the ratio of the expectations of each integral, for t in $B_{j,h}$

$$\begin{aligned}
E\widetilde{\lambda}_{n,h}(t) &= \frac{\int_{B_{j,h}} g(s)\lambda(s)\,ds}{\int_{B_{j,h}} g(s)\,ds} - E\frac{(n^{-1}\int_{B_{j,h}} J_n dM_n)\int_{B_{j,h}}(n^{-1}Y_n - g)(s)\,ds}{(\int_{B_{j,h}} g(s)\,ds)^2} \\
&\quad + E\frac{\widetilde{\lambda}_{n,h}(t)(\int_{B_{j,h}}\{n^{-1}Y_n - g)(s)\,ds\}^2}{(\int_{B_{j,h}} g(s)\,ds)^2} \\
&= \frac{\int_{B_{j,h}} g(s)\lambda(s)\,ds}{\int_{B_{j,h}} g(s)\,ds} + o(n^{-1/2}h^{1/2}) = \lambda_{n,h}(a_{j,h}) + o(h) \qquad (6.8)
\end{aligned}$$

uniformly on $\mathcal{I}_{\tau,n,h}$. The bias of $\widetilde{\lambda}_{n,h}(t)$ can be approximated by an expansion, assuming only that λ belongs to $C_1(\mathbb{R}_+)$, it is written

$$\begin{aligned}
\widetilde{b}_{\lambda,h}(t) &= \sum_{j\leq J_{\tau,h}} 1_{B_{jh}}(t)\{\lambda(a_{jh}) - \lambda(t)\} + o(h) \\
&= \sum_{j\leq J_{\tau,h}} 1_{B_{jh}}(t)|t - a_{jh}|\lambda^{(1)}(t) + o(h) = O(h)
\end{aligned}$$

also denoted $\widetilde{b}_{\lambda,h}(t) = h\widetilde{b}_{\lambda}(t) + o(h)$ and it is larger than the bias of kernel estimator. Assuming that $Var\{n^{-1/2}(Y_n - g)(t)\} = O(1)$ for every t, the variance of the denominator of $\widetilde{\lambda}_{n,h}(a_{j,h})$ equals $(nh)^{-1}Var n^{-1/2}Y_n(a_{jh}) + o((nh)^{-1}) = O((nh)^{-1}$. For every t in $B_{j,h}$, the variance of $\widetilde{\lambda}_{n,h}(t)$ is

$$\widetilde{v}_{n,h}(t) = E\{\widetilde{\lambda}_{n,h}(t) - \frac{\int_{B_{j,h}} g(s)\lambda(s)\,ds}{\int_{B_{j,h}} g(s)\,ds}\}^2 - \{E\widetilde{\lambda}_{n,h}(t) - \frac{\int_{B_{j,h}} g(s)\lambda(s)\,ds}{\int_{B_{j,h}} g(s)\,ds}\}^2,$$

and the last term is a $o(n^{-1/2}h^{1/2})$, by (6.8). Following the same calculus as for the variance of the nonparametric estimator of a regression function,

$$\begin{aligned}
\widetilde{v}_{n,h}(t) &= \{g(t) + o(h)\}^{-2}\big[Var\{(nh)^{-1}\int_{B_{j,h}} J_n\,dN_n\} \\
&\quad - 2\lambda(t)Cov\{(nh)^{-1}\int_{B_{j,h}} J_n\,dN_n, (nh)^{-1}\int_{B_{j,h}} Y_n\,ds\}\big] \\
&\quad + \lambda^2(t)Var\{(nh)^{-1}\int_{B_{j,h}} Y_n\,ds\} + o((nh)^{-1})
\end{aligned}$$

with $Var\{(nh)^{-1}\int_{B_{j,h}} J_n\, dN_n\} = (nh)^{-1}(g\lambda)(a_{jh}) + o((nh)^{-1})$ and the covariance term

$$(nh)^{-2}E\{\int_{B_{j,h}} J_n\,(dN_n - Y_n\, d\Lambda), \int_{B_{j,h}} (Y_n - g)\, ds\} = O((nh)^{-1})$$

therefore the variance of $\widetilde{\lambda}_{n,h}(t)$ can be written $\widetilde{v}_{n,h}(t) = (nh)^{-1}\widetilde{v}_\lambda(t) + o((nh)^{-1})$. The asymptotic mean squared error of the estimator $\widetilde{\lambda}_{n,h}(t)$ is minimal for the bandwidth

$$h_n(t) = n^{-1/3}\{2\widetilde{b}_\lambda^2(t)\}^{-1/3}\widetilde{v}_\lambda^{1/3}(t).$$

This expression and the AMSE do not depend on the degree of derivability of the intensity.

6.3 Risks and convergences for multiplicative intensities

The estimators (6.4) for the exponential regression of the intensity are special cases of those defined by (6.7) in a multiplicative intensity with explanatory predictable processes and an unknown regression function r. For every t in $\mathcal{I}_{n,h,\tau}$, the mean of $\widehat{\lambda}_{n,h}(t;r)$ is still $\lambda_{n,h}(t) = \int_{-1}^{1} K_h(t-s)\lambda(s)\, ds$ and their degree of derivability is the same as K.

With a parametric regression function r, the convergence in the first condition of 6.1 is replaced by the *a.s.* convergence to zero of $\sup_{t\in[0,\tau]}\sup_{\|\beta-\beta_0\|\le\varepsilon} |n^{-1}S_n^{(k)}(t;\beta) - s_0^{(k)}(t)|$, where $s_0^{(k)} = s_0^{(k)}(\beta_0)$, for $k = 0,1,2$, and $\varepsilon > 0$ is a small real number. In a nonparametric model, this condition is replaced by the *a.s.* convergence to zero of $\sup_{t\in[0,\tau]}\sup_{\|r-r_0\|\le\varepsilon} |n^{-1}S_n^{(k)}(t;r) - s_0^{(k)}(t)|$, where $s_0^{(k)} = s^{(k)}(r_0)$, for $k = 0,1,2$.

The previous conditions 6.1 are modified by the regression function. For expansions of the bias and the variance, they are now written as follows.

Condition 6.2.

(1) As n tends to infinity, the processes $n^{-1}S_n^{(k)}(t;\beta)$ and $n^{-1}S_n^{(k)}(t;r)$ are positive with a probability tending to 1 and the function defined by $s^{(k)}(t) = n^{-1}ES_n^{(k)}(t)$ belongs to class $C_2(\mathbb{R}_+)$;
(2) The function $p_n(s) = \Pr(S_n^{(0)}(s;r) > 0)$ belongs to class $C_2(\mathbb{R}_+)$ and $p_n(\tau, r_0)$ converges to 1 in probability;
(3) $\int_0^\tau r(z)g^{-1}(s)\lambda(s)\, ds < \infty$
(4) The functions λ and g belong to $C_2(\mathbb{R}_+)$ and r belongs to $C_s(\mathcal{Z})$.

6.3.1 Models with nonparametric regression functions

The regression funtion is estimated by $\widehat{r}_{n,h}(z) = \arg\max_{r_z} \widehat{\mathcal{L}}_{n,h}(z;r)$ where

$$\widehat{\mathcal{L}}_{n,h}(z;r) = n^{-1}\sum_{i=1}^{n}\int_{\mathcal{I}_{\tau,n,h}} \{\log r_z(s) + \log\widehat{\lambda}_n(s;r_z)\}\, K_h(z - Z_i(s))\, dN_i(s)$$

for t in $\mathcal{I}_{n,h,\tau}$. Its expectation $\mathcal{L}_n(z;r) = E\widehat{\mathcal{L}}_{n,h}(z;r)$ is expanded as

$$\begin{aligned}\mathcal{L}_n(z;r) &= E\int_{\mathcal{I}_{\tau,n,h}} \log\{r_z(s)\widehat{\lambda}_n(s;r_z)\}\, K_h(z - Z(s))S_n^{(0)}(s;r_Z)\lambda(s)\, ds \\ &= E\int_{\mathcal{I}_{\tau,n,h}} \log\{r_z(s)\widehat{\lambda}_n(s;r_z)\}\{S_n^{(0)}(s;r_z) + \frac{h^2}{2}\kappa_2 S_n^{(2)}(s;r_z)\} \\ &\qquad \times f_{Z(s)}(z)\lambda(s)\, ds + o(h^2),\end{aligned}$$

where $f_{Z(s)}$ is the marginal density of $Z(s)$. It follows that $\widehat{r}_n(z)$ converges uniformly, in probability, to the value $r_0(z)$ which minimizes the limit of $n^{-1}\widehat{\mathcal{L}}_n(z;r_z)$

$$\mathcal{L}_0(z;r_z) = \int_{-1}^{1} \{\log r_z(s) + \log\lambda(s;r_z)\}s^{(0)}(s;r_z)f_{Z(s)}(z)\,\lambda(s)\, ds.$$

Let $\widehat{\mathcal{L}}_n^{(k)}$ be the k-th order derivative of $\widehat{\mathcal{L}}_n(z;r)$ with respect to z, their limits are denoted $\mathcal{L}_0^{(k)}$ and their expectations $\mathcal{L}_n^{(k)}$, for $k = 1,2$.

Proposition 6.3. *Under Condition 6.2, the process $(nh)^{1/2}(\widehat{r}_{n,h} - r)$ converges weakly to a Gaussian process with mean zero and variance $(\mathcal{L}^{(2)})^{-1}V_{(1)}\{\mathcal{L}^{(2)}\}^{-1}(z;r)$ where $V_{(1)} = lim_{n\to\infty} nhVar\widehat{\mathcal{L}}_n^{(1)}$.*

Proof. The first derivative of $\widehat{\mathcal{L}}_n$ with respect to r_z and its expectation depend on the derivative of $\widehat{\lambda}_n$

$$\widehat{\lambda}_n^{(1)}(t;r_z) = -\int_{\mathcal{I}_{n,h,\tau}} K_h(t-s)S_n^{(1)}(s;r_z)S_n^{(0)-2}(s;r_z)\, d\Lambda(s),$$

$$\widehat{\mathcal{L}}_n^{(1)} = n^{-1}\sum_{i=1}^{n}\int_{\mathcal{I}_{n,h,\tau}} \{\frac{1}{r_z(s)} + \frac{\widehat{\lambda}_n^{(1)}(s;r_z)}{\widehat{\lambda}_n(s;r_z)}\} K_h(z - Z_i(s))\, dN_i(s),$$

$$\mathcal{L}_n^{(1)} = E\int_{-1}^{1}\{\frac{1}{r_{t,z}(s)} + \frac{\widehat{\lambda}_n^{(1)}(s;r_z)}{\widehat{\lambda}_n(s;r_z)}\} K_h(z - Z(s))S_n^{(0)}(s;r_z)\, f_{Z(s)}(z)\lambda(s)\, ds$$

is such that $Var\widehat{\mathcal{L}}_n^{(1)}(z;r) = O((nh)^{-1})$, therefore $(nh)^{1/2}(\widehat{\mathcal{L}}_n^{(1)} - \mathcal{L}^{(1)})(z;r)$ is bounded in probability, and the second derivative is a $O_p(1)$. By a Taylor expansion of $\widehat{\mathcal{L}}_n^{(1)}(z;r)$ in a neighbourhood of the true value $r_{0z} \equiv r_0(z)$ of the regression function at z

$$\widehat{\mathcal{L}}_n^{(1)}(z;r)-\mathcal{L}^{(1)}(z;r_0) = \{(r_z(0)-r_{0z}(s)\}^T \widehat{\mathcal{L}}_n^{(2)}(z;r_0)+O_p((\widehat{r}_{n,h}-r_0)^2(z(s)))$$

and, by an inversion, the centered estimator $(\widehat{r}_{n,h} - r_0)(z)$ is approximated by the variable $\{-\widehat{\mathcal{L}}_n^{(2)}(z;r_0)\}^{-1}\mathcal{L}^{(1)})(z;r_0)$ the variance of which is a $O((nh)^{-1})$. For every z in $\mathcal{Z}_{n,h} = \{z \in \mathcal{Z}; \sup_{z' \in \partial\mathcal{Z}} \|z - z'\| \geq h\}$, the variable $(nh)^{1/2}(\widehat{r}_{n,h} - r_0)(z)$ converges weakly to a Gaussian variable with variance $(\mathcal{L}^{(2)})^{-1} \lim_n nh_n Var \widehat{\mathcal{L}}_n^{(1)} \{\mathcal{L}^{(2)}\}^{-1}(z;r_0)$. □

Proposition 6.4. *The processes* $(nh)^{1/2}(\widehat{\lambda}_{n,h} - \lambda_{n,h})(r_0)1_{\mathcal{I}_{T,n,h}}$ *and*

$$(nh)^{1/2}(\widehat{\lambda}_{n,h}-\lambda)+(nh)^{1/2}(\widehat{r}_{n,h}-r_0)\int \frac{S_n^{(1)}}{S_n^{(0)3}}(s;\widehat{r}_{n,h})K_h(\cdot-s)\,J_n(s)\,dN_n(s)\}$$

converge weakly to the same continuous and centered Gaussian process on $\mathcal{I}_T$, *with covariances zero and variance function* $v_\lambda = \kappa_2 s^{(0)-1}(r_0)\lambda$.

Proof. The bias of $\widehat{\Lambda}_{n,h}$ is

$$\begin{aligned} b_{\Lambda,n,h}(t) &= -\int_0^T E\{S_n^{(0)}(s,r)S_n^{(0)-1}(s,\widehat{r}_{n,h}) - 1\}\,d\Lambda(s) \\ &= \int_0^T E\{(\widehat{r}_{n,h} - r)(s)\frac{S_n^{(1)}(s,\widehat{r}_{n,h}) + O_p(n^{-1/2})}{S_n^{(0)}(s,r)S,\widehat{r}_{n,h})}\}\,d\Lambda(s) \end{aligned}$$

and it is also equivalent to the mean of

$$\int_0^T (\widehat{r}_{n,h} - r)(s)S_n^{(1)}(s,\widehat{r}_{n,h})\{S_n^{(0)}(t,\widehat{r}_{n,h})\}^{-3}\,dN_n(s).$$

The bias of the estimator $\widehat{\lambda}_{n,h}$ is obtained by smoothing the bias $b_{\Lambda,n,h}(t)$ and its first order approximation can be written as the mean of

$$\widehat{b}_{\lambda,n,h}(t) = (\widehat{r}_{n,h} - r)(t)\int_{-1}^1 K_h(t-s)\frac{S_n^{(1)}(s,\widehat{r}_{n,h})}{\{S_n^{(0)}(t,\widehat{r}_{n,h})\}^3}\,dN_n(s).$$

□

6.3.2 *Models with parametric regression functions*

In the exponential regression model $r_\beta(z) = e^{\beta^T z}$ with observations $(X_i,\delta_i)_{1\leq i\leq n}$, the estimator of the parameter β minimizes $\widehat{\mathcal{L}}_n(\beta)$ which is written

$$\widehat{\mathcal{L}}_n(\beta) = n^{-1}\sum_{i=1}^n \delta_i\{\beta^T Z_i(T_i) + \log\widehat{\lambda}_n(T_i;\beta)\}.$$

Its mean

$$\mathcal{L}_n(\beta) = \int_{\mathcal{I}_{n,h,\tau}} \{\beta^T s_n^{(1)}(s;\beta_0) + E\log\{\widehat{\lambda}_n(s;\beta)\}\widehat{s}_n^{(0)}(s;\beta_0)\}\lambda(s)\,ds$$

converges to $\mathcal{L}(\beta) = \int_0^\tau \{\beta^T s^{(1)}(s;\beta_0) + \{\log\lambda(s;\beta)\}s^{(0)}(s;\beta_0)\}\lambda(s)\,ds$.

Proposition 6.5. *Under Condition 6.2, $n^{1/2}(\widehat{\beta}_{n,h} - \beta_0)$ converges weakly to a Gaussian variable with mean zero. The processes $(nh)^{1/2}(\widehat{\lambda}_{n,h} - \lambda_{n,h})(\beta_0)\,1_{\mathcal{I}_{\tau,n,h}}$ and*

$$(nh)^{1/2}(\widehat{\lambda}_{n,h}-\lambda)(\widehat{\beta}_n)+n^{1/2}(\widehat{\beta}_{n,h}-\beta_0)^T\int_{\mathcal{I}_{\tau,n,h}} \frac{S_n^{(1)}}{S_n^{(0)2}}(s;\widehat{\beta}_{n,h})K_h(\cdot-s)\,dN_n(s)$$

converge weakly to the same continuous and centered Gaussian process with covariances zero and variance function $v_\lambda = \kappa_2 s^{(0)-1}(\beta_0)\lambda$.

Proof. The derivatives with respect to β of the partial likelihood $\widehat{\mathcal{L}}_n$ are written

$$\widehat{\mathcal{L}}_n^{(1)}(\beta) = n^{-1}\sum_{i=1}^n \delta_i\{Z_i(T_i) + \frac{\widehat{\lambda}_n^{(1)}(T_i;\beta)}{\widehat{\lambda}_n(T_i;\beta)}\},$$

$$\widehat{\mathcal{L}}_n^{(2)}(\beta) = -n^{-1}\delta_i(\frac{\widehat{\lambda}_n^{(1)\otimes 2}}{\widehat{\lambda}_n^2} - \frac{\widehat{\lambda}_n^{(2)}}{\widehat{\lambda}_n})(T_i;\beta),$$

where the derivatives of $\widehat{\lambda}_{n,h}$ with respect to β are written

$$\widehat{\lambda}_n^{(1)}(t;\beta) = -n^{-1}\int_{\mathcal{I}_{\tau,n,h}} \frac{S_n^{(1)}}{S_n^{(0)2}}(s;\beta)K_h(t-s)\,dN_n(s),$$

$$\widehat{\lambda}_n^{(2)}(t;\beta) = n^{-1}\int_{\mathcal{I}_{\tau,n,h}} (2\frac{S_n^{(1)\otimes 2}}{S_n^{(0)3}} - \frac{S_n^{(2)}}{S_n^{(0)2}})(s;\beta)K_h(t-s)\,dN_n(s).$$

As h tends to zero, the predictable compensators of $\widehat{\mathcal{L}}_n^{(k)}(\beta)$, $k=1,2$, develop as

$$\mathcal{L}_n^{(1)}(\beta) = n^{-1}\int_{\mathcal{I}_{n,h,\tau}} \{S_n^{(1)}(s;\beta_0) + \frac{\widehat{\lambda}_n^{(1)}(s;\beta)}{\widehat{\lambda}_n(s;\beta)}S_n^{(0)}(s;\beta_0)\}\lambda(s)\,ds,$$

$$\mathcal{L}_n^{(2)}(\beta) = -n^{-1}\int_{\mathcal{I}_{n,h,\tau}} (\frac{\widehat{\lambda}_n^{(1)\otimes 2}}{\widehat{\lambda}_n^2} - \frac{\widehat{\lambda}_n^{(2)}}{\widehat{\lambda}_n})(s;\beta)S_n^{(0)}(s;\beta_0)\lambda(s)\,ds.$$

The expectation of $\widehat{\lambda}_n^{(1)}$, $k = 1, 2$, is deduced from the martingale property of $N_n - \widetilde{N}_n$

$$\begin{aligned}
\lambda_n^{(1)}(t;\beta) &= \int_{\mathcal{I}_{n,h,\tau}} \frac{S_n^{(1)}(s;\beta)}{S_n^{(0)2}(s;\beta)} S_n^{(0)}(s;\beta_0)\lambda(s)K_h(t-s)\,ds,\\
&= \frac{S_n^{(1)}(t;\beta)}{S_n^{(0)2}(t;\beta)} S_n^{(0)}(t;\beta_0)\lambda(t)\\
&\quad + \frac{m_{2K}h^2}{2}\{\frac{S_n^{(1)}(t;\beta)}{S_n^{(0)2}(t;\beta)} S_n^{(0)}(t;\beta_0)\lambda(t)\}^{(2)} + o(h^2),\\
\lambda_n^{(2)}(t;\beta) &= \int_{\mathcal{I}_{n,h,\tau}} (2\frac{S_n^{(1)\otimes 2}}{S_n^{(0)2}} - \frac{S_n^{(2)}}{S_n^{(0)}})(s;\beta)K_h(t-s)\lambda(s)\,ds\\
&= (2\frac{S_n^{(1)\otimes 2}}{S_n^{(0)2}} - \frac{S_n^{(2)}}{S_n^{(0)}})(t;\beta)\lambda(t)\\
&\quad + \frac{m_{2K}h^2}{2}\{(2\frac{S_n^{(1)\otimes 2}}{S_n^{(0)2}} - \frac{S_n^{(2)}}{S_n^{(0)}})(t;\beta)\lambda(t)\}^{(2)} + o(h^2).
\end{aligned}$$

It follows that $\mathcal{L}_n^{(1)}(\beta)$ and $\mathcal{L}_n^{(2)}(\beta)$ converges to $\mathcal{L}^{(1)}(t;\beta)$

$$\begin{aligned}
\mathcal{L}^{(1)}(\beta) &= \int_0^\tau \{s_n^{(1)}(s;\beta_0) - s_n^{(1)}(s;\beta)s_n^{(0)-2}(s;\beta)s_n^{(0)2}(s;\beta_0)\}\lambda(s)\,ds,\\
\mathcal{L}^{(2)}(\beta) &= -\int_0^\tau (\frac{\lambda^{(1)\otimes 2}}{\lambda^2} - \frac{\lambda^{(2)}}{\lambda})(s;\beta)s_n^{(0)}(s;\beta_0)\lambda(s)\,ds,\\
\mathcal{L}^{(2)}(\beta_0) &= -\int_0^\tau s^{(2)}(s;\beta_0)\lambda(s)\,ds,
\end{aligned}$$

where $-\mathcal{L}^{(2)}(\beta_0)$ is positive definite and $\mathcal{L}^{(1)}(t;\beta_0) = 0$ so that the maximum $\widehat{\beta}_{n,h}$ of $\widehat{\mathcal{L}}_n$ converges in probability to β_0, the maximum of the limit $\mathcal{L}$ of $\mathcal{L}_n$. The rate of convergence of $\widehat{\beta}_{n,h} - \beta_0$ is that of $\widehat{\mathcal{L}}_n^{(1)}(\beta_0)$. First $n^{1/2}(\widehat{\mathcal{L}}^{(1)})_n - \mathcal{L}^{(1)})_n)(\beta_0)$ is the sum of stochastic integrals of predictable processes with respect to centered martingales and it convergences weakly to a centered Gaussian variable with variance $v_{(1)} = \int_{\mathcal{I}_{n,h,\tau}} \{s^{(2)} - s^{(1)\otimes 2}s^{(0)-1})(s;\beta_0)\lambda(s)\,ds$. Secondly

$$\begin{aligned}
n^{1/2}(\mathcal{L}_n^{(1)} - \mathcal{L}^{(1)})(\beta_0) &= \int_{\mathcal{I}_{n,h,\tau}} [n^{1/2}\{n^{-1}(S_n^{(1)} - s^{(1)}\}(s;\beta_0)\\
&\quad + n^{1/2}\{\frac{\widehat{\lambda}_n^{(1)}}{\widehat{\lambda}_n} n^{-1}S_n^{(0)} - s^{(1)}\}(s;\beta_0)]\lambda(s)\,ds,
\end{aligned}$$

is continuous and independent of $n^{1/2}(\widehat{\mathcal{L}}_n^{(1)} - \mathcal{L}_n^{(1)})(\beta_0)$, its integrand is a sum of three terms $l_{1n} + l_{2n} + l_{3n}$ where $l_{1n} = n^{1/2}\{n^{-1}(S_n^{(1)} - s^{(1)}\}(s;\beta_0)$ convergences weakly to a centered Gaussian variable with a finite variance,

$$l_{2n} = n^{1/2}\{\frac{\lambda_{n,h}^{(1)}}{\lambda_{n,h}} n^{-1} S_n^{(0)} - s^{(1)}\}(\cdot;\beta_0),$$

$$l_{3n} = n^{1/2}[S_n^{(0)}\{\frac{\widehat{\lambda}_{n,h}^{(1)}}{\widehat{\lambda}_{n,h}} - \frac{\lambda_{n,h}^{(1)}}{\lambda_{n,h}}\}](\cdot;\beta_0).$$

The term l_{2n} convergences weakly to a centered Gaussian variable with a finite variance and l_{3n} has the same asymptotic distribution as

$$n^{1/2} s^{(0)} \lambda^{-1}\{(\widehat{\lambda}_{n,h}^{(1)} - \lambda_{n,h}^{(1)}) - \lambda^{(1)}\lambda^{-1}(\widehat{\lambda}_{n,h} - \lambda_{n,h})\}(\cdot;\beta_0)$$

where the process

$$(nh)^{1/2}(\widehat{\lambda}_{n,h} - \lambda_{n,h})(t;\beta_0)$$
$$= n^{1/2}\int_0^T S_n^{(0)-1}(s;\beta_0) K_h(t-s)\, Y_n(s)\, d(N_n - \widetilde{N}_n)(s)$$

has the mean zero and the variance $h\int_{\mathcal{I}_{\tau,n,h}} S_n^{(0)-1}(s;\beta_0)K_h^2(t-s)\lambda(s)\,ds$ which converges in probability to $v_\lambda = \kappa_2 s^{(0)-1}(t;\beta_0)\lambda(t)$. In the same way, the process

$$n^{1/2}(\widehat{\lambda}_{n,h}^{(1)} - \lambda_{n,h}^{(1)})(t;\beta_0) = n^{1/2}\int_{\mathcal{I}_{n,h,\tau}} \frac{S_n^{(1)}}{S_n^{(0)2}}(s;\beta_0)K_h(t-s)\, d(N_n - \widetilde{N}_n)(s)$$

is consistent and it has the finite asymptotic variance

$$v_{\lambda,(1)}(t) = s^{(1)\otimes 2} s^{(0)-3}(t;\beta_0)\lambda(t).$$

The term l_{3n} with asymptotic variance zero converges in probability to zero. The proof of the weak convergence of $\widehat{\beta}_n$ ends as previously. The process $(nh)^{1/2}(\widehat{\lambda}_{n,h} - \lambda)(t;\widehat{\beta}_n)$ develops as

$$n^{1/2}\int_{\mathcal{I}_{\tau,n,h}} J_n(s) S_n^{(0)-1}(s;\widehat{\beta}_n)K_h(t-s)\, d(N_n - \widetilde{N}_n)(s)$$
$$+ \int_{\mathcal{I}_{\tau,n,h}} \{S_n^{(0)-1}(s;\widehat{\beta}_n) - S_n^{(0)-1}(s;\beta_0)\} S_n^{(0)}(s;\beta_0)K_h(t-s)\lambda(s)\,ds$$

the first term of the right-hand side converges weakly to a centered Gaussian process with variance $\kappa_2 s^{(0)-1}(t;\beta_0)\lambda(t)$ and covariances zero, and the second term is expanded into

$$-\, n^{1/2}(\widehat{\beta}_{n,h} - \beta_0)^T \int_{\mathcal{I}_{\tau,n,h}} S_n^{(1)}(s;\beta_0) S_n^{(0)-1}(s;\beta_0)K_h(t-s)\lambda(s)\,ds$$
$$= -n^{1/2}(\widehat{\beta}_{n,h} - \beta_0)^T s^{(1)}(t;\beta_0) s^{(0)-1}(t;\beta_0)\lambda(t) + o(1). \qquad \square$$

The results are analogous for every parametric regression function r_β of C_2, the processes are then defined by $S_n^{(0)}(t;\beta) = \sum_{i=1}^n r_{Z_i}(t;\beta)1_{\{T_i\geq t\}}$ and $S_n^{(k)}$ is its derivative of order k with respect to β

$$\widehat{\mathcal{L}}_n^{(1)}(\beta) = n^{-1}\sum_{i=1}^n \delta_i\{\frac{r_\beta^{(1)}(Z_i(T_i))}{r_\beta(Z_i(T_i))} + \frac{\widehat{\lambda}_{n,h}^{(1)}(s;\beta)}{\widehat{\lambda}_n(s;\beta)}\}\,dN_n(s),$$

$$\mathcal{L}_n^{(1)}(\beta) = E\int_{\mathcal{I}_{n,h,\tau}}\{\frac{r_\beta^{(1)}(Z_i(t)}{r_\beta(Z_i(t)} + \frac{\widehat{\lambda}_n^{(1)}(s;\beta)}{\widehat{\lambda}_n(s;\beta)}\} \times S_n^{(0)}(t;\beta_0)\lambda(s)\,ds,$$

$$\widehat{\mathcal{L}}_n^{(2)}(\beta) = -n^{-1}\sum_{i=1}^n \delta_i(\frac{\widehat{\lambda}_n^{(1)2}}{\widehat{\lambda}_n^2} - \frac{\widehat{\lambda}_n^{(2)}}{\widehat{\lambda}_n})(T_i;\beta)\,.$$

All results of this section are extended to varying bandwidth estimators as before.

6.4 Histograms for intensity and regression functions

The histogram estimator (6.2) for the intensity λ with a parametric regression or (6.6), for nonparametric regression of the intensity, are consistent estimators as h tends to zero and n to infinity. Their expectations are approximated like (6.8) by a ratio of means. Their variances are calculated as in Section 6.2.2.

The nonparametric regression function r is estimated by

$$\widetilde{r}_{n,h}(z) = \sum_{l\leq L_h} \widetilde{r}_{n,h}(z_{lh})1_{D_{lh}}(z)$$

and the histogram estimator for the intensity defines the estimator $\widetilde{r}_{n,h}$ of the regression function by

$$\widetilde{r}_{n,h}(z_{lh}) = \arg\max_{r_l}\prod_{T_i\leq\tau}[\{\sum_{l\leq L_h} r_l 1_{D_{lh}}(Z_n(T_i))\}\widetilde{\lambda}_{n,h}(T_i;r_{lh})]^{\delta_i},$$

$$\widetilde{\lambda}_{n,h}(t;r) = \sum_{j\leq J_h} 1_{B_{jh}}(t)\int_{B_{jh}} J_n(s)\,dN_n(s)\,[\int_{B_{jh}} S_n^{(0)}(s;r)\,ds]^{-1}.$$

For every t in $B_{j,h}$, let $S_n^{(0)}(t;r) = \sum_{i=1}^n r_{Z_i}(t)Y_i(t)$, the limit of $n^{-1}S_n^{(0)}(t;r)$ is $s^{(0)}(t;r) = \sum_{l\leq L_h} r_{z_{lh}}\Pr(Z(t)\in D_{lh}) + o(1)$ and its variance is

$$v^{(0)}(t;r) = n^{-1}\sum_{l\leq L_h} r_{z_{lh}}^2[\Pr(Z(t)\in D_{lh})g(t) - \{\Pr(Z(t)\in D_{lh})g(t)\}^2] + o(1)$$

under Condition 6.1. The mean of $\widetilde{\lambda}_{n,h}(t; z_{lh})$ is

$$\lambda_{n,h}(t; r_{lh}) = \frac{\int_{B_{j,h}} s^{(0)}(s; r_{lh})\lambda(s)\,ds}{\int_{B_{j,h}} s^{(0)}(s; r_{lh})\,ds} + O((nh)^{-1}) = \lambda(a_{j,h}) + o(h)$$

and its bias is $\widetilde{b}_{\lambda,h}(t; r_{lh}) = h\lambda^{(1)}(t) + o(h)$, uniformly on $\mathcal{I}_{\tau,n,h}$. Its variance is

$$\begin{aligned}
\widetilde{v}_{n,h}(t; r_{lh}) &= E\{\widetilde{\lambda}_{n,h}(t) - \frac{\int_{B_{j,h}} s^{(0)}(s; r_{lh})\lambda(s)\,ds}{\int_{B_{j,h}} s^{(0)}(s; r_{lh})\,ds}\}^2 + O((nh)^{-1}) \\
&= \{s^{(0)}(t; r_{lh})\}^{-2}\big[Var\{(nh)^{-1}\int_{B_{j,h}} J_n(s)\,dN_n\} \\
&\quad - 2\lambda(t)Cov\{(nh)^{-1}\int_{B_{j,h}} J_n\,dN_n, (nh)^{-1}\int_{B_{j,h}} S_n^{(0)}(s; r_{lh})\,ds\}\big] \\
&\quad + \lambda^2(t)Var\{(nh)^{-1}\int_{B_{j,h}} S_n^{(0)}(s; r_{lh})\,ds\} + o((nh)^{-1}) + o(h)
\end{aligned}$$

where, for t in $B_{j,h}$ and $Z(t)$ in D_{lh}

$$\begin{aligned}
&Var\{(nh)^{-1}\int_{B_{j,h}} J_n\,dN_n\} = (nh)^{-1}s^{(0)}(t; r_{lh})\lambda(t) + o((nh)^{-1}) \\
&Var\{(nh)^{-1}\int_{B_{j,h}} S_n^{(0)}(s; r_{lh})\,ds\} = (nh)^{-1}v^{(0)}(t; r_{lh}) + o((nh)^{-1}), \\
&\mathrm{Cov}\{(nh)^{-1}\int_{B_{j,h}} J_n\,dN_n, (nh)^{-1}\int_{B_{j,h}} S_n^{(0)}(s; r_{lh})\,ds\} \\
&\qquad\qquad = (nh)^{-1}v^{(0)}(t; r_{lh})\lambda(t) + o((nh)^{-1})
\end{aligned}$$

therefore $\widetilde{v}_{n,h}(t) = (nh)^{-1}\widetilde{v}_\lambda(t) + o((nh)^{-1})$ with

$$\widetilde{v}_\lambda(t) = s^{(0)-2}(s; r_{lh})\{s^{(0)}(t; r_{lh})\lambda(t) - v^{(0)}(t; r_{lh})\lambda^2(t)\}$$

and $(nh)^{1/2}(\widetilde{\lambda}_{n,h} - \lambda_{n,h})(t)$ converges weakly to a centered Gaussian process with variance $\widetilde{v}_\lambda(t)$.

The asymptotic mean squared error of the estimator $\widetilde{\lambda}_{n,h}(t)$ is still minimal for the bandwidth $h_n(t) = n^{-1/3}\{2\widetilde{b}_\lambda^2(t)\}^{-1/3}\widetilde{v}_\lambda^{1/3}(t)$. The stepwise constant estimator of the nonparametric regression function r maximizes

$$\begin{aligned}
\mathcal{L}_{n,h}(r) = \int\{\sum_{l\le L_h} \log r_l 1_{D_{lh}}(Z_n(s))\}\,J_n(s)\,dN_n(s) \\
+ \int \widetilde{\lambda}_{nh}(s; r_{lh})\,J_n(s)\,dN_n(s)
\end{aligned}$$

and it satisfies $\mathcal{L}^{(1)}_{n,h}(\widetilde{r}_{1h},\ldots,\widetilde{r}_{L_h,h}) = 0$, where $\mathcal{L}^{(1)}_{n,h}$ is a vector with components the derivatives with respect to the components of $r_h = (r_{lh})_{l\leq L_h}$

$$\mathcal{L}^{(1)}_{n,h,l}(r_h) = \int \{\sum_{l\leq L_h} \frac{1}{r_{lh}} 1_{D_{lh}}(Z_n(s))\} J_n(s)\, dN_n(s) + \int \widetilde{\lambda}^{(1)}_{n,h,l}(s; r_{lh})\, J_n(s)\, dN_n(s).$$

The derivatives of the intensity are consistently estimated by differences of values of the histogram, in the same way as the derivatives of a density. The variance of $\widetilde{\lambda}^{(1)}_{n,h}$ is a $O((nh^3)^{-1})$ and the estimator of the regression function converges with that rate.

In the parametric regression model, the histogram estimator for the function λ and the related estimator of the regression parameter have the same form (6.5), where the function r and the process $S^{(0)}_n$ are indexed by the parameter β. Let t in $B_{j,h}$

$$\widetilde{\lambda}_{n,h}(t;\beta) = \sum_{j\leq J_h} 1_{B_{jh}}(t)[\int_{B_{jh}} J_n(s)\, dN_n(s)][\int_{B_{jh}} S^{(0)}_n(s;\beta)\, ds]^{-1},$$

$$\widetilde{\beta}_{n,h} = \arg\max_{\beta} \prod_{T_i\leq\tau} \{r_{Z_i}(T_i;\beta)\widetilde{\lambda}_{n,h}(T_i;\beta)\}^{\delta_i}.$$

The derivative of $\widetilde{\lambda}_{n,h}$ are obtained by deriving $S^{(0)}_n$ with respect to β

$$\widetilde{\lambda}^{(1)}_{n,h}(s;\beta) = -\sum_{j\leq J_h} 1_{B_{jh}}(t)[\int_{B_{jh}} J_n(s)\, dN_n(s)] \frac{\int_{B_{jh}} S^{(1)}_n(s;\beta)\, ds}{[\int_{B_{jh}} S^{(0)}_n(s;\beta)\, ds]^2}$$

and the derivative of the logarithm of the partial likelihood for β is

$$\mathcal{L}^{(1)}_{n,h}(\beta) = n^{-1}\sum_{i=1}^{n} \int \frac{r^{(1)}_{Z_i}}{r_{Z_i}}(s;\beta)\, dN_i(s) + n^{-1}\int \widetilde{\lambda}^{(1)}_{nh}(s;\beta)\, dN_n(s).$$

It is zero at $\widetilde{\beta}_{n,h}$ and its convergence rate is a $O((nh^3))$, like $\widetilde{\lambda}^{(1)}_{nh}$. Therefore, $\widetilde{\beta}_{n,h}$ has the convergence rate $O((nh^3))$ and the estimator of the hazard function has the convergence rate $O((nh))$.

6.5 Estimation of the density of duration excess

For the indicator N_T of a time variable T, the probability of excess is

$$\begin{aligned} P_t(t+x) &= \Pr(T > t+x \mid T > t) = 1 - \Pr(t < T \leq t+x \mid T > t) \\ &= 1 - \Pr\{N_T(t+x) - N_T(t) = 1 \mid N_T(t) = 0\}. \end{aligned}$$

For a sample of n independent and identically distributed variables $(T_i)_{i\leq n}$, the processes $n^{-1}N_n(t) = n^{-1}\sum_{i=1}^n 1_{T_i\leq t}$ and $n^{-1}Y_n(t^-) = 1 - n^{-1}N_n(t)$ converge respectively to the functions $F(t)$ and $\bar{F}(t)$, and $P_t(t+x)$ is estimated by $\widehat{P}_{n,t}(t+x) = 1 - \{N_n(t+x) - N_n(t)\}1_{\{N_n(t)<n\}}\{n - N_n(t)\}^{-1}$. Let $T_{1:n} = \min_{1\leq i\leq n} T_i$ and $T_{n:n} = \max_{1\leq i\leq n} T_i$, on every interval $[a,b]$ included in $I_n =]T_{1:n}, T_{n:n}[$, the product-limit estimator has values in the open interval $]0,1[$ and the centered process is approximated as

$$\begin{aligned}n^{1/2}\{P_t(t+x) - \widehat{P}_{n,t}(t+x)\} &= n^{1/2}\bar{F}^{-1}(t)[n^{-1}\{N_n(t+x) - N_n(t)\} \\ &- F(t+x) + F(t) - P_t(t+x)\{n^{-1}Y_n(t^-) - \bar{F}(t)\}] + o_{L_2}(1)\end{aligned}$$

and it converges weakly to a Gausian process with mean zero and a finite variance. The probability $P_t(t+x) = \exp\{-\int_t^{t+x}\lambda(s)\,ds\}$ is also estimated by the product-limit estimator on the interval $]T_{1:n}, T_{n:n}[$

$$\widehat{P}_{n,t}(t+x) = \prod_{1\leq i\leq n}\{1 - \frac{1_{\{t<T_i\leq t+x\}}J_n(T_i)}{Y_n(T_i)}\}, \tag{6.9}$$

with $Y_n = 1_{\{J_n>0\}}$. The estimator is constant between jump times T_i and the size of the jumps is $\Delta\widehat{P}_{n,t}(T_i) = 1_{\{t<T_i\}}\widehat{P}_{n,t}(T_i^-)J_n(T_i)Y_n^{-1}(T_i)$. According to the asymptotic results established for the product-limit estimator, for every interval $[a,b]$ included in the interval $]T_{1:n}, T_{n:n}[$, the product-limit estimator has values in the open interval $]0,1[$ and $\sup_{(t,x+t)\in[a,b]}|\widehat{P}_{n,t}(t+x) - P_t(t+x)|$ converges *a.s.* to zero. Let

$$B_{n,t}(t+x) = n^{1/2}\frac{P_t - \widehat{P}_{n,t}}{P_t}(t+x) \tag{6.10}$$

also written $\widehat{P}_{n,t}(t+x) = P_t(t+x)\{1 - n^{-1/2}B_{n,t}(t+x)\}$.

Theorem 6.2. *On every interval $[a,b]$ included in the interval $]T_{1:n}, T_{n:n}[$, the estimator $\widehat{P}_n$ satisfies*

$$B_{n,t}(t+x) = n^{1/2}\int_t^{t+x} d(\widehat{\Lambda}_n - \Lambda).$$

It converges weakly to a Gaussian process B with independent increments, mean zero and a finite variance $v_B(t+x;t) = \int_t^{t+x}\bar{F}^{-1}\,dF$.

Proof. Let $[a,b]$ be a sub-interval of $]T_{1:n}, T_{n:n}[$ and let $\widehat{\Lambda}_n$ be the martingale estimator of $\Lambda = -\log\bar{F}$, $\widehat{\Lambda}_n = \int_0^t J_nY_n^{-1}\,dN_n$ is uniformly consistent on $[a,b]$ and $n^{1/2}(\widehat{\Lambda}_n - \Lambda)$ converges weakly to a centered Gaussian process with independent increments and variance $v_\Lambda(t) = \int_0^t \bar{F}^{-2}\, dF$. The process $B_{n,t}(t+x)$ is expanded as

$$
\begin{aligned}
-P_t(t+x)B_{n,t}(t+x) &= n^{1/2}[\exp\{-\int_t^{t+x} d\widehat{\Lambda}_n\} - \exp\{-\int_t^{t+x} d\Lambda\}] \\
&\quad + n^{1/2}[\exp\{\log \widehat{P}_{n,t}(t+x)\} - \exp\{-\int_t^{t+x} d\widehat{\Lambda}_n\}] \\
&= -n^{1/2} \int_t^{t+x} d(\widehat{\Lambda}_n - \Lambda) \exp\{-\int_t^{t+x} d\Lambda_n^*\} \{1 + o(1)\} \\
&\quad + n^{1/2} \exp\{-\int_t^{t+x} d\Lambda_n^{**}\} \{\log \frac{\widehat{\bar{F}}_n(t+x)}{\widehat{\bar{F}}_n(t)} - \int_t^{t+x} d\widehat{\Lambda}_n\},
\end{aligned}
$$

where Λ_n^* is between Λ and $\widehat{\Lambda}_n$, so it converges uniformly to Λ, and $-\int_t^{t+x} \Lambda_n^{**}$ is between $\log \widehat{\bar{F}}_n(t+x)\} - \log \widehat{\bar{F}}_n(t)$ and $\int_t^{t+x} d\widehat{\Lambda}_n$. From Lemma 1.5.1 in Pons (2008), the variable $n^{1/2} \sup_{[a,b]} |-\log\{\widehat{\bar{F}}_n(t)\} - \widehat{\Lambda}_n(t)|$ converges in probability to zero as n tends to infinity then the last term of the expansion of $B_{n,t}(t+x)$ converges to zero in probability, uniformly on $[a,b]$. The first term converges weakly to $-P_t(t+x)\, B_\Lambda(t, t+x)$ where the process $B_\Lambda(t, t+x)$ is the limiting distribution of $n^{1/2} \int_t^{t+x} d(\widehat{\Lambda}_n - \Lambda)$. □

By Equation (6.10), the covariance of $n^{1/2}(\widehat{P}_{n,t} - P_t)(t+x)$ and $n^{1/2}(\widehat{P}_{n,t} - P_t)(t+y)$ is denoted

$$
\begin{aligned}
C_P(t+x, t+y; t) &= P_t(t+x)P_t(t+y) \lim_n EB^2_{n,t}(t + x \wedge y) \\
&= P_t(t+x)P_t(t+y)v_B(t + x \wedge y; t).
\end{aligned}
$$

The weak convergence of $n^{1/2}(\bar{F}_n - \bar{F})$ on the interval $[0, T_{n:n}]$ allows to extend the previous proposition to a weak convergence on $[T_{1:n}, T_{n:n}]$. For $t < \tau_F$, if $\int_0^{\tau_F} \bar{F}^{-1}\, d\Lambda < \infty$

$$
\frac{P_t - \widehat{P}_{n,t}}{P_t}(t+x) = \int_{t \vee T_{1:n}}^{(t+x)\wedge T_{n:n}} \frac{1 - \widehat{F}_n(s^-)}{1 - F(s)}\, d(\widehat{\Lambda}_n - \Lambda)(s). \tag{6.11}
$$

Therefore, the process defined for t and $t+x$ in $[T_{1:n}, T_{n:n}]$ by $n^{1/2}\{(P_t - \widehat{P}_{n,t})P_t^{-1}\}(t+x)$ converges weakly on the support of F to a centered Gaussian process B_P, with independent increments and variance $v_{\bar{F}}(t+x) - v_{\bar{F}}(t)$. By the product definition of the estimator of the probability $P_t(t+x)$, it satisfies

$$
\widehat{P}_{n,t}(t+x) = \int_{t \vee T_{1:n}}^{(t+x)\wedge T_{n:n}} \widehat{P}_{n,t}(t+s^-)\, d\widehat{\Lambda}_n(s).
$$

The estimator is extended without modification to samples of right-censored variables, with an independent censorship by a sequence of independent and identically distributed variables with survival function $\bar{G}$. Only the definitions of N_n and Y_n and the expression of the variances are modified by a multiplicative term $\bar{G}^{-1}$ in the integral defining v_Λ, as in the variance of the classical product-limit estimator. For $p \geq 2$, by Equation (6.11), their exists a constant c_p such that

$$E\{P_t(t+x) - \widehat{P}_{n,t}(t+x)\}^p \leq c_p P_t(t+x)^p E\Big[\int_t^{t+x} \left\{\frac{\widehat{\bar{F}}_n(s^-)}{\bar{F}(s)}\right\}^2 \frac{dN_n(s)}{Y_n^2(s)}\Big]^{p/2}$$

therefore $\|P_t(t+x) - \widehat{P}_{n,t}(t+x)\|_p = O(n^{-1/2})$.

The probability density of excess duration is the derivative of the probability $P_t(t+x)$, $p_t(t+x) = -f(t+x)/\bar{F}(t)$. A kernel estimator of $p_t(t+x)$ is now defined, for $t < t + x - h_n$ as

$$\begin{aligned}
\widehat{p}_{n,h}(t+x;t) &= \int_{-1}^{1} K_h(t+x-s)\, d\widehat{P}_{n,t}(s) \\
&= \int_{-1}^{1} K_h(t+x-s)\widehat{P}_{n,t}(s^-)\, d\widehat{\Lambda}_n(s) \\
&= \sum_{i=1}^{n} K_h(T_i - t - x)\widehat{P}_{n,t}(T_i^-) J_n(T_i) Y_n^{-1}(T_i)\delta_i 1_{\{t<T_i\}} \\
&= \sum_{i=1}^{n} K_h(T_i - t - x)\delta_i 1_{\{t<T_i\}} P_t(T_i) \frac{\widehat{\bar{F}}_n(T_i^-)}{\bar{F}(T_i)} \frac{J_n}{Y_n}(T_i).
\end{aligned}$$

As in Section 2.8, the estimator $\widehat{p}_{n,h}$ is uniformly consistent. For a density p in class C_2, its bias deduced from (6.11) is $b_{p,n,h}(t+x;t) = \frac{h^2}{2} p_t(t+x) + o(h^2)$. Its variance has the bound

$$\begin{aligned}
v_{p,n,h}(t+x;t) \leq 2[E \int K_h^2(t+x-s) p_t(s)^2 \Big\{\int_0^s \frac{\widehat{\bar{F}}_n^-}{\bar{F}}\, d(\widehat{\Lambda}_n - \Lambda)\Big\}^2 ds \\
+ E \int K_h^2(t+x-s) \frac{\widehat{\bar{F}}_n^2(s^-)}{\bar{F}(t)} Y_n^{-1}(s)\, d\Lambda(s)] = O((nh)^{-1}).
\end{aligned}$$

It follows that for every t and x, $(nh)^{1/2}(\widehat{p}_{n,h} - p_{n,h})(t+x;t)$ converges weakly to a Gaussian variable with mean zero and finite variance $v_p(t+x;t) = \lim_n nhv_{p,n,h}(t+x;t)$. The covariance of $\widehat{p}_{n,h}(t+x;t)$ and $\widehat{p}_{n,h}(t+y;t)$ is written

$$C_{n,h,p}(t+x,t+y;t) = \int\int K_h(t+x-u)K_h(t+y-v) Cov\{\widehat{P}_{n,t}(du)\, \widehat{P}_{n,t}(dy)\}.$$

For $x < y$, the covariance of $\widehat{P}_{n,t}(x)$ and $\widehat{P}_{n,t}(y)$ is $n^{-1}P_t(t+x)P_t(t+y)v_B(t+x;t)+o(n^{-1})$, hence for $|x-y| > 2h$, the limit of $nC_{n,h,p}$ converges to

$$\int\int K_h(t+x-u)K_h(t+y-v)p_t(t+u)p_t(t+v)v_B(t+u;t) + P_t(t+u)P_t(t+v)v_B^{(1)}(t+u;t)\} \, du \, dv$$
$$= p_t^{(1)}(t+y)\{p_t(t+x)v_B(t+x;t)\}^{(1)} + p_t(t+y)\{P_t(t+x)v_B^{(1)}(t+x;t)\}^{(1)},$$

if $|x-y| \le 2h$, $C_{n,h,p} = 0((nh)^{-1})$. Then, the process $(nh)^{1/2}(\widehat{p}_{n,h} - p)$ converges weakly to a Gaussian variable with mean zero, covariances zero and variance function v_p.

6.6 Estimators for processes on increasing intervals

Let $(N(t))_{t\le T}$ be the counting processes associated to a sequence of random time variables $(T_i)_{i\ge 1}$

$$N_T(t) = \sum_{i\ge 1} 1_{\{T_i \le t\wedge T\}}, t \ge 0,$$

where T is deterministic or a random stopping time. Its predictable compensator is written $\widetilde{N}_T(t) = \sum_{i=1}^{N(T)} \int_0^t Y_i(s)\mu(s, Z_i(s)) \, ds$ as in the previous sections. In a model without covariates, $Y_T(t) = \sum_{i=1}^{N(T)} Y_i(t)$ and the baseline intensity λ of the intensity $\mu_T(t) = \lambda(t)Y_T(t)$ is estimated for t in $[h, T-h]$ by smoothing the estimator of the cumulative hazard function Λ, $\widehat{\Lambda}_T(t) = \int_0^{t\wedge T} J_T(s)Y_T^{-1}(s) \, d\widetilde{N}_T(s)$, with an indication $J_T(t) = 1_{\{Y_T(t)>0\}}$

$$\widehat{\lambda}_{T,h_T}(t) = \int_{-1}^{1} Y_T^{-1}(s)J_T(s)K_h(t-s) \, dN_T(s). \tag{6.12}$$

The mean of $\widehat{\lambda}_{T,h_T}$ is $\lambda_{T,h_T}(t) = \int_{-1}^{1} E\{Y_T^{-1}(s)\}K_h(t-s) \, d\Lambda(s)$. Conditions for its consistency are an ergodic property for the process (N_T, Y_T) and Conditions 6.1 written for Y_T with limits as T tends to infinity, then its bias develops as $b_{\lambda,T,h_T}(t) = \frac{h_T^s}{s!} m_{sK}\lambda^{(s)}(t) + o(h_T^s)$. The variance of the estimator is $Var\{\widehat{\lambda}_{T,h_T}(t)\} = (Th_T)^{-1}\kappa_2 \, g^{-1}(t)\lambda(t) + o((Th_T)^{-1})$ and the L_p-norms of the estimator satisfy $\sup_{t\in\mathcal{I}_{T,h,\tau}} \|\widehat{\lambda}_{T,h_T}(t) - \lambda_{T,h_T}(t)\|_p = 0((Th_T)^{-1/p})$ and $\sup_{t\in\mathcal{I}_{T,h_T,\tau}} \|\widehat{\lambda}_{T,h_T}(t) - \lambda(t)\|_p = 0((Th_T)^{-1/p} + h^s)$. Under a mixing condition for the point process (N_T, Y_T) which ensures the weak convergence of the process $T^{1/2}(Y_T - g)$, the process $(Th_T)^{1/2}(\widehat{\lambda}_{T,h_T}(t) - \lambda)$

converges weakly to the Gaussian process limit of Theorem 6.1. Now the optimal bandwidth are $O(T^{-1/(2s+1)})$ and the minimal mean squared errors are $O(T^{-2s/(2s+1)})$.

In models with covariates, the process Y_T is modified by a regression function at the jump times as $S_T^{(0)}(t;\beta) = \sum_{i=1}^{N(T)} r_{Z_i}(t;\beta)1_{\{T_i \geq t\}}$ with a parametric regression function or $S_T^{(0)}(t;r) = \sum_{i=1}^{N(T)} r_{Z_i}(t;)1_{\{T_i \geq t\}}$ with a nonparametric regression function. The predictable compensator of N_T becomes $\widetilde{N}_T(t) = \int_0^{t\wedge T} S_T^{(0)}(s;r)\lambda(s)\,ds$. The function λ is estimated by smoothing the estimator of the cumulative hazard function

$$\widehat{\lambda}_{T,h}(t;\beta) = \int_{-1}^{1} \{S_T^{(0)}(s;\beta)\}^{-1} K_h(t-s)\,dN_T(s)$$

and the regression function is estimated using the parameter estimator $\widehat{\beta}_{T,h_T} = \arg\max_\beta \prod_{T_i \leq T}\{r_{Z_i}(t;\beta)\widehat{\lambda}_{T,h_T}(T_i;\beta)\}^{\delta_i}$ or by

$$\begin{aligned}\widehat{r}_{T,h_T}(z) = \arg\max_{r_z} \sum_{i=1}^{N(T)} \int_0^T & K_{h_2}(z - Z_i(s))\{\log r_z(s) \\ & + \log \widehat{\lambda}_{T,h_T}(s;r_z)\} K_{h_2}(z - Z_i(s))\,dN_i(s)\end{aligned}$$

in the nonparametric case.

Assuming that the process $Z((T_i))_{i\geq 1}$ is bounded, ergodic and has finite moments up to order 4, the processes $T^{-1}S_T^{(k)}(s;\beta)$ and $T^{-1}S_T^{(k)}(s;r)$ converge in probabilty and uniformly on $[0,T]$ and in a neighbourhood of β_0 or r_0 to finite limits $s^{(k)}$ and $T^{1/2}(T^{-1}S_T^{(k)} - s^{(k)})$ satisfies a theorem central limit, for $k = 0,1,2$. All results for the kernel estimators of Section 6.3 are still valid under these conditions replacing n by T.

The empirical estimator of the probability of excess duration and a kernel estimator for its density are defined by

$$\begin{aligned}\widehat{P}_{T,t}(t+x) &= \prod_{t \leq s \leq T\wedge(t+x)} \{1 - J_T(s)Y_T^{-1}(s)\,dN_T(s)\}, \\ \widehat{p}_{T,h_T}(t+x;t) &= \int_{-1}^{1} K_h(t+x-s)\,d\widehat{P}_{T,t}(s).\end{aligned}$$

The process $B_{T,t}(t+x) = T^{1/2}\{P_t(t+x) - \widehat{P}_{T,t}(t+x)\}P_t^{-1}(t+x)$ satisfies

$$B_{T,t}(t+x) = \int_t^{t+x} \frac{\widehat{\bar{F}}_T^-}{\bar{F}}\, d(\widehat{\Lambda}_T - \Lambda).$$

6.7 Models with varying intensity or regression coefficients

More complex models are required to describe the distribution of event times when the conditions may change in time or according to the value of a variable. Pons (1999, 2002) presented results for two extensions of the classical exponential regression model for an intensity involving a non-parametric baseline hazard function and a regression on a p-dimensional process Z

- a model for the duration $X = T^0 - S$ of a phenomenon starting at S and ending at T^0, with a non-stationary baseline hazard depending non-parametrically on the time S at which the observed phenomenon starts

$$\lambda_{X|S,Z}(x \mid S, Z) = \lambda_{X|S}(x; S)\, e^{\beta^T Z(S+x)}, \tag{6.13}$$

- a model where the regression coefficients are smooth functions of an observed variable X

$$\lambda(t \mid X, Z) = \lambda(t) e^{\beta(X)^T Z(t)}. \tag{6.14}$$

The asymptotic properties of the estimators $\widehat{\beta}_n$ and $\widehat{\Lambda}_n$ of β and of the cumulative baseline hazard function follow the classical lines but the kernel smoothing of the likelihood requires modifications.

In model (6.13), the time T^0 may be right-censored at a random time C independent of (S, T^0) conditionally on Z and non informative for the parameters β and $\lambda_{X,S}$, and that S is uncensored. We observe a sample $(S_i, T_i, \delta_i, Z_i)_{1 \le i \le n}$ drawn from the distribution of (S, T, δ, Z), where $T = T^0 \wedge C$ and $\delta = 1_{\{T^0 \le C\}}$ is the censoring indicator. The data are observed on a finite time interval $[0, \tau]$ strictly included in the support of the distributions of the variables S, T^0 and C, and (S, T^0) belongs to the triangle $I_\tau = \{(s, x) \in [0, \tau] \times [0, \tau]; s + x \le \tau\}$. For S_i in a neighborhood of s, the baseline hazard $\lambda_{X|S}(\cdot; S_i)$ is approximated by $\lambda_{X|S}(\cdot; s)$, which yields a local log-likelihood at $s \in [h_n, \tau - h_n]$, defined as

$$\begin{aligned} l_n(s) &= \sum_i K_{h_n}(s - S_i)\delta_i\{\log \lambda_{X|S}(X_i; S_i) + \beta^T Z_i(T_i)\} \\ &\quad - \int_0^\tau Y_i(y) \exp\{\beta^T Z_i(S_i + y)\}\lambda_{X|S}(y; S_i)\, dy. \end{aligned}$$

Let $\widetilde{X}_i = X_i \wedge (C_i - S_i)$

$$I_{n,\tau} = \{(s,x); s \in [h_n, \tau - h_n], x \in [0, \tau - s]\},$$
$$Y_i(x) = 1_{\{T_i^0 \wedge C_i \geq S_i + x\}} = 1_{\{\widetilde{X}_i \geq x\}},$$
$$S_n^{(0)}(x; s, \beta) = n^{-1} \sum_j K_{h_n}(s - S_j) Y_j(x) \exp\{\beta^T Z_j(S_j + x)\}.$$

The estimator of $\Lambda_{0,X|S}(x; s) = \int_0^x \lambda_{0,X|S}(y; s)\, dy$ is defined for $(s, x) \in I_{n,\tau}$ by $\hat{\Lambda}_{n,X|S}(x; s) = \hat{\Lambda}_{n,X|S}(x; s, \hat{\beta}_n)$ with

$$\hat{\Lambda}_{n,X|S}(x; s, \beta) = \sum_i \frac{K_{h_n}(s - S_i) 1_{\{S_i \leq C_i, X_i \leq x \wedge (C_i - S_i)\}}}{n S_n^{(0)}(X_i; s, \beta)}.$$

The estimator $\widehat{\beta}_n$ of the regression coefficient maximizes the following partial likelihood

$$l_n(\beta) = \sum_i \delta_i \left[\beta^T Z_i(T_i^0) - \log\{n S_n^{(0)}(X_i; S_i, \beta)\}\right] \varepsilon_n(S_i)$$

where $\varepsilon_n(s) = 1_{[h_n, \tau - h_n]}(s)$. The bandwidth h is supposed to converge to zero, with nh^2 tends to infinity and $h = o(n^{-1/4})$ as n tends to infinity, the other conditions are precised by Pons (2002). The variable $n^{1/2}(\widehat{\beta}_n - \beta_0)$ converges weakly to a Gaussian variable $\mathcal{N}(0, I^{-1}(\beta_0))$ where the variance $I^{-1}(\beta_0)$, defined as the inverse of the limit of the second derivative of the partial likelihood l_n, is the minimal variance for a regular estimator of β_0.

The weak convergence of the estimated cumulative hazard function defined along the current time and the duration elapsed between two events relies on the bivariate empirical processes

$$\widehat{H}_n(s, x) = n^{-1} \sum_i \delta_i 1_{\{S_i \leq s\}} 1_{\{X_i \leq x\}},$$
$$\bar{W}_n^{(0)}(s, x) = n^{-1} \sum_i e^{\beta_0^T Z_i(S_i + x)} 1_{\{S_i \leq s\}} 1_{\{\widetilde{X}_i \geq x\}},$$
$$\widehat{B}_n = n^{1/2}(\bar{W}_n^{(0)} - W^{(0)}, \widehat{H}_n - H)\, 1_{I_{n,\tau}},$$

under boundedness and regularity conditions, the process $\widehat{B}_n$ converges weakly to a Gaussian limit. With functions λ and $s^{(0)}$ in class C_2, the bias of the estimator $\widehat{\Lambda}_{n,X|S}$ is a $O(h^2)$, thus the optimal bandwidth minimizing the asymptotic mean squared error of Λ is $O(n^{-1/5})$ and it is still written in terms of the squared bias and the variance of the estimator. If the regressor Z is a bounded variable, then there exists a sequence of centered Gaussian processes B_n on $I_{n,\tau}$ such that $\|\widehat{B}_n - B_n\|_{I_{n,\tau}} = o_p(h_n^{1/2})$. This property implies the weak convergence of the process $(nh_n)^{1/2}(\widehat{\Lambda}_{n,X|S} - \Lambda_{0,X|S})1_{\{I_{n,\tau}\}}$ to a centered Gaussian process.

In model (6.14), $\widehat{\Lambda}_n$ only involves kernel terms through the regression functions but both $\widehat{\beta}_n$ and $\widehat{\Lambda}_n$ have the same non-parametric rate of convergence. In Pons (1999), the estimator $\widehat{\beta}_{n,h}(x)$ was defined as the value of β which maximizes

$$l_{n,x}(\beta) = \sum_{i\leq n} \delta_i K_{h_n}(x - X_i)[\{\beta(X_i)\}^T Z_i(T_i) \tag{6.15}$$

$$- \log\{\sum_{j\leq n} K_{h_n}(x - X_j) Y_j(T_i) e^{\{\beta(X_i)\}^T Z_j(T_i)}\}] \tag{6.16}$$

where $Y_i(t) = 1_{\{T_i \geq t\}}$ is the risk indicator for individual i at t. Let $S_n^{(0)}(t,\beta) = \sum_i Y_i(t) e^{\beta(X_i)^T Z_i(t)}$, an estimator of the integrated baseline hazard function is $\widehat{\Lambda}_n(t) = \int_0^t S_n^{(0)-1}(s, \widehat{\beta}_{n,h})\, dN_n(s)$. For every x in $\mathcal{I}_{X,h}$, the process $n^{-1} l_{n,x}$ converges uniformly to

$$l_x(\beta) = \int_0^\tau (\beta - \beta_0)(x)^T s^{(1)}(t, \beta_0(x), x)$$

$$- s^{(0)}(t, \beta_0(x), x) \log \frac{s^{(0)}(t, \beta(x), x)}{s^{(0)}(t, \beta_0(x), x)}\, d\Lambda_0(t)$$

which is maximum at β_0 hence $\widehat{\beta}_{n,h}(x) = \arg\max l_{n,x}(x)$ converges to $\beta_0(x)$. Let $U_{n,h}(\cdot, x)$ and $I_{n,h}(\cdot, x)$ be the first two derivatives of the process $l_{n,x}$ with respect to β at fixed x in $\mathcal{I}_{X,h}$, the estimator of $\beta(x)$ satisfies $U_{n,h}(\widehat{\beta}_{n,h}(x), x) = 0$ and $I_{n,h}(x) \leq 0$ converges uniformly to a limit $I(x)$. By a Taylor expansion $U_{n,h}(\beta_0(x), x) = (\widehat{\beta}_{n,h}(x) - \beta_0(x))^T \{I(\beta_0, x) + o(1)\}$ and $(\widehat{\beta}_{n,h}(x) - \beta_0(x)) = \{I_{n,h}(\beta_0, x) + o(1)\}\}^{-1} U_{n,h}(\beta_0(x)$. Under the assumptions that the bandwidth is a $O(n^{-1/5})$ and the function β belongs to the class $C_2(\mathcal{I}_X)$, the bias of $\widehat{\beta}_{n,h}(x)$ is approximated by $I^{-1}(\beta_0, x) h^2 u(x)$ where $u(x)$ has the form $u(x) = \frac{m_{2K}}{2} \int_0^\tau \phi(t,x)\, d\Lambda_0(t)$ and its variance is $(nh)^{-1}\kappa_2 I^{-1}(\beta_0, x) + o((nh)^{-1})$. The asymptotic mean integrated squared error $AMISE_w(h) = E \int_{\mathcal{X}_{n,h}} \|\widehat{\beta}_{n,h}(x) - \beta_0(x)\| w(x)\, dx$ for $\widehat{\beta}_{n,h}(x)$ is therefore minimal for the bandwidth

$$h_{n,opt} = n^{-1/5} \frac{\kappa_2 \int_{\mathcal{X}_{n,h}} \|I^{-1}(\beta_0, x)\| w(x)\, dx}{\int_{\mathcal{X}_{n,h}} u(x) \|I^{-1}(\beta_0, x)\| w(x)\, dx}$$

and the error $AMISE_w(h_{n,opt})$ has the order $n^{-2/5}$.

The limiting distributions of the estimators are now expressed in the following proposition. Let $G_n = (n^{-1}h)^{1/2}\{S_n^{(0)}(\widehat{\beta}_{n,h}) - S_n^{(0)}(\beta_0)\}$.

Proposition 6.6. *For every x in $\mathcal{I}_{X,n,h}$, the variable $(nh_n)^{1/2}(\widehat{\beta}_{n,h} - \beta_0)(x)$ converges weakly to a Gaussian variable $\mathcal{N}(0, \gamma_2(K)I_0^{-1}(x))$.*

The process $(nh_n)^{1/2}(\widehat{\Lambda}_n - \Lambda_0)$ converges weakly to the Gaussian process $-\int_0^{\cdot} G(t)\{\int s^{(0)}(t,y)\,dy\}^{-2}\,d\Lambda_0(t)$, where the process G is the limiting distribution of G_n.

The convergence rate $(nh_n)^{1/2}$ for the estimator of Λ comes from the variance $E\int_0^t S_n^{(0)}(s,\beta_0)S_n^{(0)-2}(s,\widehat{\beta}_{n,h})\,d\Lambda_0(s)$ of $\widehat{\Lambda}_n(t) - \Lambda_0(t)$ developed by a first order Taylor expansion.

6.8 Progressive censoring of a random time sequence

Let $(T_i)_{i=1,\dots,n}$ be a sequence of independent random time variables and $(T_j)_{j=1,\dots,m}$ be an independent sequence of independent random censoring times such that a random number R_j of variables T_i are censored at T_j and $\sum_{j=1}^m R_j = n$. Then the censored variables $X_{i,j} = T_i \wedge C_j$ are no longer independent, only m sets of variables are independent. Let

$$N_{n,m}(t) = \sum_{j=1}^m \sum_{i=1}^{R_j} 1_{\{T_i \le t \wedge C_j\}}, \; Y_{n,m}(t) = \sum_{j=1}^m \sum_{i=1}^{R_j} 1_{\{X_{i,j} \ge t\}}.$$

Let F be the common distribution function of the variables T_i and G_j be the distribution function of C_j, the intensity of the point process $N_{n,m}$ is still written $\lambda_{n,m}(t) = \lambda Y_{n,m}$ with $\lambda = \bar{F}^{-1}f$. Conditionally on the censoring number $R = (R_1, \dots, R_m)$, the expectations of $N_{n,m}(t)$ and $Y_{n,m}(t)$ are

$$E\{N_{n,m}(t) \mid R\} = \sum_{j=1}^m R_j \int_0^t \bar{G}_j\,dF,$$

$$E\{Y_{n,m}(t) \mid R\} = \sum_{j=1}^m R_j \bar{G}_j(t)\bar{F}(t).$$

Let $\mu_R = \lim R_j$ for $j = 1, \dots, m$, and $J_{n,m}(t) = 1_{\{Y_{n,m}(t)>0\}}$, the estimator of the cumulative hazard function Λ and its derivative λ are

$$\widehat{\Lambda}_{n,m}(t) = \int_0^t Y_{n,m}^{-1} J_{n,m}\,dN_{n,m},$$

$$\widehat{\lambda}_{n,m}(t) = \int K_h(t-s)\,d\widehat{\Lambda}_{n,m}(s).$$

Assuming that there exists an uniform limit for the mean survival function $\bar{G} = \lim_{m\to\infty} m^{-1}\sum_{j=1}^{m}\bar{G}_j$, the process $n^{-1}Y_{n,m}$ converges uniformly to its expectation $\mu_Y(t) = \mu_R\bar{G}(t)\bar{F}(t)$, and $n^{-1}N_{n,m}$ converges uniformly to $\mu_R\int_0^t \bar{G}\,dF$. The estimators $\widehat{\Lambda}_{n,m}$ and $\widehat{\lambda}_{n,m}$ are then unbiased and uniformly consistent. The variance of $n^{1/2}(\widehat{\Lambda}_{n,m}-\Lambda)(t)$ is $\int_0^t nY_{n,m}^{-1}J_{n,m}\,d\Lambda$ and it converges to $v_\Lambda(t) = \int_0^t \mu_Y^{-1}\,d\Lambda$. The variance of the estimator $\widehat{\lambda}_{n,m}(t)$ is $(nh)^{-1}\kappa_2 v_\Lambda^{(1)}(t) + o((nh)^{-1})$ and both estimator processes converge weakly to Gausian processes with zero mean and these variances. The process $n^{1/2}(\widehat{\Lambda}_{n,m}-\Lambda)$ has independent increments and $n^{1/2}(\widehat{\lambda}_{n,m}-\lambda)$ has asymptotic covariances zero. All results for multiplicative regression models with independent censoring times apply to this progressive random censoring scheme. With nonrandom numbers R_j, the necessary condition for the convergence of the processes is the uniform convergence of $n^{-1}\sum_{j=1}^{m} R_j\bar{G}_j$.

6.9 Exercises

(1) Define a kernel estimator for a Poisson process N with a functional intensity $\lambda(t)$ from the observation of a sample-path on an interval $[0,T]$, such that T tends to infinity and prove that the process $(Th)^{-1/2}(\widehat{\lambda}_{T,h}-\lambda)$ converges weakly.

(2) Calculate the bias of the estimator of β which maximizes the process $l_{n,x}$ defined by (6.15).

(3) Consider a point process with a conditional multiplicative intensity $\lambda(t)$ $r(\beta^T Z(t))\ Y(t)$, where λ and r are nonparametric real functions and β is a vector of unkown parameters. Define kernel estimators for the functions λ and r and an estimator for β by minimization of a partial likelihood. Determine the order of their risks and prove their weak convergence.

Chapter 7

Estimation in semi-parametric regression models

7.1 Introduction

In the single-index regression model, the scalar response variable Y is expressed as a nonparametric transform for a linear combination of the components of a vector X of d regression variables

$$Y_i = g(\theta^T X_i) + \sigma(\eta^T X_i)\varepsilon_i, \tag{7.1}$$

where X is a vector of regression variables in a bounded subset of $\mathbb{R}^d$ and ε is an error variable such that $E(\varepsilon|X) = 0$ and $Var(\varepsilon \mid X = x) = 1$, then $Var(Y \mid X = x) = \sigma^2(\eta^T x)$. The model includes a similar parametrization for the mean and the variance functions. The parameters are vectors η and θ belonging to an open and bounded subset Θ of $\mathbb{R}^d$ and g, an unknown function of $C_2(\mathbb{R})$. Several estimators for the semi-parametric regression function $m(x) = g(\theta^T x)$ have been defined from approximations and they are calculated by iterations, without model for the variance. Here the estimators are defined in a two-step procedure from the weighted estimator of the regression function g defined by (3.19). The true parameter value is a vector in $\mathbb{R}^{2d}$

$$(\eta_0^T, \theta_0^T)^T = \arg\min_{\eta,\theta\in\Theta} V(\eta, \theta) \tag{7.2}$$

where

$$V(\eta, \theta) = E[\sigma^{-1}(\eta^T X_i)\{Y - g(\theta^T X)\}^2]$$

is the mean weighted squared error at a fixed parameters value η and θ. Several empirical criteria can be defined for estimating V, estimators of θ_0 satisfying the same property (7.2) are deduced.

Let $(X_i, Y_i)_{i=1,\dots,n}$ be a sample of observations in $\mathbb{R}^{d+1}$, in the model with known variance function. At fixed θ, let $\widehat{g}_{n,h}(z)$ be the nonparametric regression estimator defined with regressors values $\theta^T X_i$ in a neighborhood of z, then the parameter θ is estimated by minimizing a mean squared error of estimation, which is a goodness-of-fit criterium for the model. For observations such that $\theta^T X_i$ lies in a neighborhood of $z = \theta^T x$, the regression function is estimated at fixed η and θ, by

$$\widehat{g}_{n,h}(z;\eta,\theta) = \frac{\sum_{i=1}^n \widehat{\sigma}_{n,h}^{-1}(\eta^T X_i) Y_i K_h(z-\theta^T X_i)}{\sum_{i=1}^n \widehat{\sigma}_{n,h}^{-1}(\eta^T X_i) K_h(z-\theta^T X_i)},$$

and $\widehat{\sigma}_{n,h}^{-1}(\eta^T X_i)$ is the estimator (3.19) calculated at fixed η. The global goodness-of-fit error and the estimator of θ minimizing this error are

$$\widehat{V}_{n,h}(\eta,\theta) = n^{-1}\sum_{i=1}^n \widehat{\sigma}_{n,h}^{-1}(\eta^T X_i)\{Y_i - \widehat{g}_{n,h}(\theta^T X_i;\theta)\}^2, \tag{7.3}$$

$$(\widehat{\eta}_{n,h}^T, \widehat{\theta}_{n,h}^T)^T = \arg\min_{\eta,\theta\in\Theta} \widehat{V}_{n,h}(\eta,\theta), \tag{7.4}$$

$$\widehat{m}_{n,h}(x) = \widehat{g}_{n,h}(\widehat{\theta}_{n,h}^T x; \widehat{\theta}_{n,h}).$$

We first assume that the variance function is known and denoted $\sigma^2(x)$, the error and the estimator of g are then only normalized by $\sigma^{-1}(x)$. The global error (7.3) and the estimator (7.4) have been modified by considering the mean of local empirical squared errors. In a neighborhood of z, a local empirical squared error is defined by smoothing (7.3)

$$\widehat{V}_{n,h}(z;\theta) = n^{-1}\sum_{i=1}^n \sigma^{-1}(X_i)\{Y_i - \widehat{g}_{n,h}(\theta^T X_i;\theta)\}^2 K_h(z-\theta^T X_i),$$

$$\widehat{\theta}_{n,h,z} = \arg\min_{\theta\in\Theta} \widehat{V}_{n,h}(z;\theta).$$

A global estimator $\bar{\theta}_{n,h}^T$ of θ was defined by an empirical mean of the local estimators at the random point $\widehat{Z}_{n,i} = \widehat{\theta}_{n,h,Z_i}^T Z_i$. Then an estimator of the regression function m is

$$\bar{m}_{n,h}(x) = \widehat{g}_{n,h}(\bar{\theta}_{n,h}^T x; \bar{\theta}_{n,h}). \tag{7.5}$$

Another estimator is then obtained by minimizing the differential criterion

$$\widehat{W}_{n,h}(\theta) = n^{-2}\sum_{i\neq j=1}^n \{\sigma^2(X_i)+\sigma^2(X_j)\}^{-1/2}\{Y_i - Y_j - \widehat{g}_{n,h}(\theta^T X_i;\theta) + \widehat{g}_{n,h}(\theta^T X_j;\theta)\}^2 K_h(X_i - X_j), \tag{7.6}$$

$$\widetilde{\theta}_{n,h} = \arg\min_{\theta\in\Theta} \widehat{W}_{n,h}(\theta),$$

with sums on values individuals having close values of the regression variable. Finally, a third estimator of the regression function is

$$\widetilde{m}_{n,h}(x) = \widehat{g}_{n,h}(\widetilde{\theta}_{n,h}^T x; \widetilde{\theta}_{n,h})$$

with the estimator (7.6) for the vector parameter θ.

The error $V(\theta_0)$ and $W(\theta_0)$ is estimated by plug-in with (7.4) and

$$\widehat{V}_{n,h} = \widehat{V}_{n,h}(\widehat{\theta}_{n,h})$$
$$\widehat{W}_{n,h} = \widehat{W}_{n,h}(\widehat{\theta}_{n,h})$$

and $\bar{V}_{n,h} = n^{-1}\sum_{k=1}^n \widehat{V}_{n,h}(\bar{\theta}_{n,h}^T X_k; \bar{\theta}_{n,h})$ for an empirical mean of local estimators.

Various forms of the estimation criteria are weighted mean square errors in order to take into account the unknown variance σ^2 of the variable Y or other weighting functions. Iterative procedures were also defined with alternative estimations of the function g and the parameters. In the next section, the convergence rate of the parameter estimator and the regression function $\widehat{m}_{n,h}$ are determined for (7.3) and (7.6) with variance 1. In the presence of nonparametric estimator of the function g, the convergence rate of the parameter estimator differs from the parametric rate. The limiting distributions of the parametric and nonparametric estimators are studied for the global errors $\widehat{V}_{n,h}$ and $\widehat{W}_{n,h}$.

More general models are nonparametric regressions with a parametric change of variable. In the nonparametric regression with a change of variables, the linear expression $\theta^T X$ is replaced by a transformation of X using a parametric family of functions defined in $\mathbb{R}^d$, $\phi = \{\varphi_\theta\}_{\theta\in\Theta}$ subset of the class C_2. The semi-parametric regression model is

$$Y = g \circ \varphi_\theta(X) + \sigma(X)\varepsilon. \tag{7.7}$$

The error is still $V(\theta, \sigma) = E\sigma^{-1}(X)\{Y - m(X)\}^2$ with $m = g \circ \varphi_\theta$, at fixed θ and σ, and the parameters can be estimated by minimizing over θ and σ similar empirical estimators of $V(\theta)$ as above.

7.2 Convergence of the estimators

Proposition 2.3 and Theorem 2.1 imply the convergence in probability to zero of the empirical error criteria $\sup_{\theta\in\Theta} |\widehat{V}_{n,h}(\theta) - V(\theta)|$ and $\sup_{\theta\in\Theta} |\widehat{W}_{n,h}(\theta) - W(\theta)|$, the local criterium $\widehat{V}_{n,h}(z;\theta)$ also converges uniformly on $\mathcal{Z} \times \Theta$ to $V(\theta; z) = E[\{Y - m(X)\}^2 | \theta^T X = z]$. As the minimum of the limits $V(\theta)$ and $W(\theta)$ is θ_0, all estimators converge to θ_0. The minimization of $\widehat{V}_{n,h}$ provides an estimator $\widehat{\theta}_n$ solution of $\widehat{V}^{(1)}_{n,h}(\theta)=0$ where

$$\widehat{V}^{(1)}_{n,h}(\theta)=2n^{-1}\sum_{i=1}^{n}\{Y_i-\widehat{g}_{n,h}(\theta^TX_i;\theta)\}\{\widehat{g}^{(1)}_{n,h}(\theta^TX_i;\theta)\}X_i$$

and for the second derivative, let $Z_i=\theta^TX_i$, at fixed θ

$$\widehat{V}^{(2)}_{n,h}(\theta)=2n^{-1}\sum_{i=1}^{n}[\{Y_i-\widehat{g}_{n,h}(Z_i;\theta)\}\widehat{g}^{(2)}_{n,h}(Z_i;\theta)-\{\widehat{g}^{(1)}_{n,h}(Z_i;\theta)\}^2]X_i^{\otimes 2}.$$

The limit of $\widehat{V}^{(2)}_{n,h}(\theta)$ is the second derivative of $V(\theta)=E[\{g(\theta^TX)-g(\theta_0^TX)\}^2]$ which is minimal at θ_0, therefore $-V^{(2)}_{n,h}(\theta)$ is positive definite in a neighbourhood of θ_0 for n large enough since that is true for the limiting function $-V^{(2)}(\theta)$. Expanding $\widehat{V}^{(1)}_{n,h}(\theta)$, for θ in a neighbourhood of θ_0, implies $\widehat{V}^{(1)}_{n,h}(\theta)=\widehat{V}^{(1)}_{n,h}(\theta_0)+(\theta-\theta_0)^T\widehat{V}^{(2)}_{n,h}(\theta_0)+O_P((\theta-\theta_0)^2)$. Then the estimators satisfy $\widehat{\theta}_{n,h}-\theta_0=\{-\widehat{V}^{(2)}_{n,h}(\theta_0)\}^{-1}\widehat{V}^{(1)}_{n,h}(\theta_0)+O_P((\widehat{\theta}_{n,h}-\theta_0)^2)$ also written

$$\widehat{\theta}_{n,h}-\theta_0=\{-\widehat{V}^{(2)}_{n,h}(\theta_0)\}^{-1}\{\widehat{V}^{(1)}_{n,h}(\theta_0)-V^{(1)}(\theta_0)\}+O_P((\widehat{\theta}_{n,h}-\theta_0)^2).$$

Applying Theorem 2.1, $(nh^3)^{1/2}\{\widehat{g}^{(1)}_{n,h}(\theta_0)-g^{(1)}(\theta_0)\}$ converges weakly to a continuous Gaussian process with zero mean and finite variance. The first two moments of $\widehat{V}^{(1)}_{n,h}$ are $V^{(1)}_{n,h}(\theta)=E\widehat{V}^{(1)}_{n,h}(\theta)$, expanded at θ_0 as

$$\begin{aligned}V^{(1)}_{n,h}(\theta_0)&=2E[\{m(X_i)-g_{n,h}(\theta_0^TX_i;\theta_0)\}\{\widehat{g}^{(1)}_{n,h}(\theta_0^TX_i;\theta_0)\}X_i]\\&\quad-2E[(\widehat{g}_{n,h}-g_{n,h})(\theta_0^TX_i;\theta_0)\}\{\widehat{g}^{(1)}_{n,h}(\theta_0^TX_i;\theta_0)\}X_i].\end{aligned}$$

Proposition 10.1 in Appendix A and Appendix C prove that $V^{(1)}_{n,h}(\theta_0)$ is a $O(h^2)+O((nh^2)^{-1})$, and the variance of $\widehat{V}^{(1)}_{n,h}(\theta_0^Tx;\theta_0)$ equals $(nh^3)^{-1}\Sigma_{V,\theta_0}+o((nh^3)^{-1})$, where

$$\Sigma_{V,\theta_0}=E[X^{\otimes 2}f_X^{-2}(X)\{m_g^2(\theta_0^TX)+\sigma_g^2(\theta_0^TX)\}w_2(\theta_0^TX)](\int K^{(1)2}).$$

It follows that the variable $(nh^3)^{1/2}\{\widehat{V}^{(1)}_{n,h}(\theta_0)-V^{(1)}_{n,h}(\theta_0)\}$ converges weakly to a continuous Gaussian variable with mean zero and variance Σ_{V,θ_0}, and

$$(nh^3)^{1/2}(\widehat{\theta}_{n,h}-\theta_0)=2\{\widehat{V}^{(2)}_{n,h}(\theta_0)\}^{-1}(nh^3)^{1/2})\{V^{(1)}-\widehat{V}^{(1)}_{n,h}\}(\theta_0)+o_p(1). \tag{7.8}$$

Moreover, the second derivative $\widehat{V}^{(2)}_{n,h}$ converges uniformly to a bounded function $V^{(2)}_\theta$ on the parameter space. The result extend to regression functions of $C_s(\mathcal{I}_X)$, $s \geq 2$, and the order nh^3 related to the variance of the derivative of $\widehat{g}_{n,h}$ is unchanged. Let $v_0 = V_{\theta_0}^{(2)-1}\Sigma_{V,\theta_0}V_{\theta_0}^{(2)-1}$.

Proposition 7.1. *Under Conditions 2.1 and 3.1, and if $h_n = o(n^{-1/7})$, the estimators of the parameter θ and the function m in class $C_s(\mathcal{I}_X)$ are consistent, $(nh^3)^{1/2}(\widehat{\theta}_{n,h} - \theta_0)$ converges weakly to a Gaussian variable with with mean zero and variance v_0 and $(nh^3)^{1/2}(\widehat{m}_{n,h} - m)$ converges weakly to a Gaussian process with mean zero and covariance function $g^{(1)}(\theta_0^T x)v_{\theta_0}g^{(1)}(\theta_0^T x')$ at (x, x').*

Proof. The process $\widehat{m}_{n,h}(x) - m(x)$ splits as the sum of two terms

$$u_{n,h}(x) = \widehat{g}_{n,h}(\widehat{\theta}_{n,h}^T x) - g(\widehat{\theta}_{n,h}^T x),$$
$$v_{n,h}(x) = g(\widehat{\theta}_{n,h}^T x) - g(\theta_0^T x).$$

The convergence rate of $v_{n,h}$ is the same as $\widehat{\theta}_{n,h} - \theta$, which is $(nh^3)^{1/2}$ like $(\widehat{V}_{n,h}^{(1)} - V^{(1)})(\theta_0)$ since the bias of $\widehat{V}_{n,h}^{(1)}(\theta_0)$ disappears with $h_n = o(n^{-1/7})$. The process $(nh)^{1/2}u_{n,h}(x) = (nh)^{1/2}(\widehat{g}_{n,h} - g)(\theta_0^T x)\{1 + o_p(1)\}$ is a $O_p(1)$, then the convergence rate of $\{\widehat{m}_{n,h}(x) - m(x)\}$ is $(nh^3)^{1/2}$. □

The bandwidth minimizing the sum of the squared bias and the variance of $\widehat{V}_{n,h}^{(1)}(\theta_0)$ is $h_{V,n} = O(n^{-1/7})$ and the convergence rate of $\widehat{V}_{n,h}^{(1)}(\theta_0)$ is $n^{2/7}$ in that case. Note that with a bandwidth $h_n = O(n^{-1/7})$, the limit is a Gaussian variable with the finite mean $\lim_n (nh^7)^{1/2} V_{n,h}^{(2)-1})(\theta_0)V_{n,h}^{(1)}(\theta_0)$. This rate $h_n = O(n^{-1/7})$ is optimal for estimating the first derivative of a function of class C_2 and the biases were obtained under this assumption. This is a consequence of the approximation of $\widehat{\theta}_{n,h} - \theta_0$ in terms of $(\widehat{V}_{n,h}^{(1)} - V^{(1)})(\theta)$. The optimal local and global bandwidths for the estimators $\widehat{g}_{n,h}$ and $\widehat{m}_{n,h}$ are $O(n^{-1/(2s+3)})$ and they are expressed as a ration of their (integrated) bias and variance, the mean squared errors for their estimation are $O(n^{-2s/(2s+3)})$.

The estimator minimizing the local error $\widehat{V}_{n,h}(z;\theta)$ is approximated by linearization of the error for $\theta^T X_i$ in a neighbourhood of z

$$\widehat{V}_{n,h}(\theta; z) = n^{-1}\sum_{i=1}^{n} K_h(z - \theta^T X_i)\{Y_i - \widehat{g}_{n,h}(z) - (\theta^T X_i - z)\widehat{g}_{n,h}^{(1)}(z)\}^2 + O(h^2),$$

and the derivatives of the linear approximation are considered with variables $Z_i = \theta^T X_i$ in a neighborhood of z. The asymptotic behaviour of its derivatives differs from those of the global error due to the kernel smoothing. The mean estimator has a smaller variance than the estimator minimizing the global error $\widehat{V}_{n,h}$.

For independent variables such that $X_i \neq X_j$ and $|X_i - X_j| \leq h$, let $\Delta_{i,j}X = X_i - X_j$ and $\Delta_{i,j}\varphi(X) = \varphi(X_i) - \varphi(X_j)$, for every function φ. Let $Z = \theta^T X$ at fixed θ. With the notations of Proposition 3.1, the differential criterion $\widehat{W}_{n,h}$ defined by (7.6) has the mean

$$\begin{aligned} W_{n,h}(\theta) &= E[\{\Delta_{i,j}(Y - \widehat{g}_{n,h})(\theta^T X)\}^2 K_h(X_i - X_j)] \\ &= E[\{\sigma^2(\theta^T X_i) - \sigma^2(\theta^T X_j)\} + \{g(\theta_0^T X_i) - g(\theta_0^T X_j) - g(\theta^T X_i) \\ &\quad + g(\theta^T X_j)\}^2 + E[(nh)^{-1}\{\sigma_g^2(\theta^T X_i) - \sigma_g^2(\theta^T X_j)\} \\ &\quad + h^4\{b_{g,n,h}^2(\theta^T X_i) - b_{g,n,h}^2(\theta^T X_j)\} K_h(X_i - X_j)]\{1 + o(1)\}, \end{aligned}$$

where $\widehat{g}_{n,h}$ is the regression function estimator. By Lemma 10.1 in Appendix C, it is approximated by $O(h^2) + O(n^{-1}h)$. The first derivative of $\widehat{W}_{n,h}$ is

$$\begin{aligned} \widehat{W}_{n,h}^{(1)}(\theta) = -2n^{-2} \sum_{i \neq j = 1}^{n} &\{Y_i - Y_j - \widehat{g}_{n,h}(Z_i;\theta) + \widehat{g}_{n,h}(Z_j;\theta)\} \\ &\{\widehat{g}_{n,h}^{(1)}(Z_i;\theta)X_i - \widehat{g}_{n,h}^{(1)}(Z_j;\theta)X_j\} K_h(\Delta_{i,j}X) \end{aligned}$$

its mean develops as

$$\begin{aligned} W_{n,h}^{(1)}(\theta) &= 2E[\{\Delta_{i,j}(m_0(X) - g_{n,h}(Z))\}\{\Delta_{i,j}(Xg_{n,h}^{(1)}(Z))\} K_h(\Delta_{i,j}X)] \\ &\quad - E[\{\Delta_{i,j}(\widehat{g}_{n,h} - g_{n,h})(Z)\}\{\Delta_{i,j}(X\widehat{g}_{n,h}^{(1)}(Z))\} K_h(\Delta_{i,j}X)], \end{aligned}$$

denoted $E_1 + E_2$. The first term is expanded as

$$E_1 = h^2 m_{2K} \int \{m_0(x) - g_{n,h}(\theta^T x)\}^{(1)}\{xg_{n,h}^{(1)}(\theta^T x)\}^{(1)} f_X(x)\, dF_X(x),$$

using a first order approximation for the variations, E_2 depends on the covariance of $\widehat{g}_{n,h}^{(2)}(Z)$ and $\widehat{g}_{n,h}^{(1)}(Z)$, with a factor h^2, thus the second term is a $0(n^{-1}h^{-2})$. The function $W_{n,h}^{(1)}(\theta)$ is then an uniform $O(h^2) + O((nh^2)^{-1})$ and it tends to zero uniformly on the bounded parameter space. As in Appendix C, the main term of its variance $\Sigma_{W,n,h}(\theta_0)$ depends on $E\{\widehat{g}_{n,h}^{(2)}(Z) - g_{n,h}^{(2)}(Z)\}^2\{\widehat{g}_{n,h}^{(1)}(Z) - g_{n,h}^{(1)}(Z)\}^2 = O(nh^4)^{-1}$, with a factor $n^{-2}h^3$, thus it is a $0(n^{-3}h^{-1})$. The second order derivative is an empirical mean

$$\begin{aligned} \widehat{W}_{n,h}^{(2)}(\theta) = 2n^{-2} \sum_{i \neq j = 1}^{n} &[\{\Delta_{i,j}\{X\widehat{g}_{n,h}^{(1)}(\theta^T X)\}\}^{\otimes 2} - \{Y_i - Y_j - \Delta_{i,j}\widehat{g}_{n,h}(\theta^T X)\} \\ &\times \Delta_{i,j}\{X^{\otimes 2}\widehat{g}_{n,h}^{(2)}(\theta^T X)\}] K_h(\Delta_{i,j}X), \end{aligned}$$

it is approximated by its expectation

$$\begin{aligned}
W_{n,h}^{(2)} &= 2E[\Delta_{i,j}^{\otimes 2}\{X\widehat{g}_{n,h}^{(1)}(\theta^T X)\} - \{\Delta_{i,j}\{m_0(X) - \widehat{g}_{n,h}(\theta^T X)\} \\
&\qquad \times \Delta_{i,j}\{X^{\otimes 2}\widehat{g}_{n,h}^{(2)}(\theta^T X)\}]\, K_h(\Delta_{i,j}X) \\
&= 2E\, K_h(\Delta_{i,j}X)\, [\{\Delta_{i,j}^{\otimes 2}\{Xg^{(1)}(\theta^T X)\} \\
&\quad - \Delta_{i,j}\{m_0(X) - g_{n,h}(\theta^T X)\}\Delta_{i,j}\{X^{\otimes 2}g_{n,h}^{(2)}(\theta^T X)\} \\
&\quad + \Delta_{i,j}^{\otimes 2}\{(\widehat{g}_{n,h} - g_{n,h})(\theta^T X)\}\Delta_{i,j}\{X^{\otimes 2}\widehat{g}_{n,h}^{(2)}(\theta^T X)\}].
\end{aligned}$$

The sequence $(W_{n,h_n}^{(2)})_n$ is therefore an uniform $O(h^2 + (nh^3)^{-1})$. Assuming that nh_n^4 tends to infinity with n, $W_{n,h_n}^{(1)} = O(h^2 + (nh^2)^{-1}) = O(h^2)$ and a necessary condition for the weak convergence of $(n^3h)^{1/2}\widehat{W}_{n,h}^{(1)}$ is that its normalized mean converges, *i.e.* $h = O(n^{-3/5})$.

Under this condition, nh^5 tends to zero and the convergence rate of $W_{n,h_n}^{(2)}$ is h^2. Arguing as for the estimator of $\widehat{\theta}_{n,h}$ related to $\widehat{V}_{n,h}$, $\widetilde{\theta}_{n,h}$ minimizing $\widehat{W}_{n,h}$ is such that $(n^3h^5)^{1/2}(\widetilde{\theta}_{n,h} - \theta_{n,h})$ is approximated by the variable $h^2\{\widehat{W}_{n,h}^{(2)}(\theta_0)\}^{-1}n^{3/2}h^{1/2}(\widehat{W}_{n,h}^{(1)} - W_{n,h}^{(1)})(\theta_0)$. It converges weakly to a Gaussian process with variance $v_{\theta_0} = W^{(2)-1}(\theta_0)\Sigma_{W,\theta_0}W^{(2)-1}(\theta_0)$. Following the arguments for the proof of the previous proposition, we obtain the following convergences.

Proposition 7.2. *Under Conditions 2.1 and 3.1, and if* $h_n = O(n^{-3/5})$, *the estimators of the parameter* θ *and of the function* m *in class* C_2 *are consistent,* $(n^3h^5)^{1/2}(\widetilde{\theta}_{n,h} - \theta_0)$ *and* $(nh)^{1/2}(\widetilde{m}_{n,h} - m)$ *converge weakly to Gaussian processes with finite variances.*

With $h_n = O(n^{-3/5})$, the limit of $(nh)^{1/2}(\widetilde{m}_{n,h} - m)$ is centered, if moreover $h_n = o(n^{-3/5})$, both limits are centered. The weighted estimator of the regression function has the same convergence rate as in the model with a constant variance as proved in Section 3.6, the convergence rates of the estimators of θ and g of Propositions 7.1 and 7.2 in the single-index model are therefore unchanged.

7.3 Nonparametric regression with a change of variables

In the semi-parametric regression model (7.7), with $m = g \circ \varphi_\theta$, the estimators are built following the same lines as in the single index regression model. For observations such that $\varphi_\theta(X_i)$ lies in a neighbourhood of $z = \varphi_\theta(x)$, the regression function is estimated by

$$\widehat{g}_{n,h}(z;\theta) = \frac{\sum_{i=1}^n \widehat{\sigma}_{n,h}^{-1}(X_i) Y_i K_h(z - \varphi_\theta(X_i))}{\sum_{i=1}^n \widehat{\sigma}_{n,h}^{-1}(X_i) K_h(z - \varphi_\theta(X_i))}$$

with a parametric or semi-parametric estimator for the variance function. The global goodness-of-fit error and the estimator of θ minimizing this error are

$$\widehat{V}_{n,h}(\theta) = n^{-1} \sum_{i=1}^n \widehat{\sigma}_{n,h}^{-1}(X_i) \{Y_i - \widehat{g}_{n,h} \circ \varphi_\theta(X_i)\}^2,$$

$$\widehat{\theta}_n = \arg\min_{\theta \in \Theta} V_{n,h}(\theta),$$

and $\widehat{m}_{n,h} = \widehat{g}_{n,h} \circ \varphi_{\widehat{\theta}_{n,h}}$.

Assume that the variance is constant and let $Z_i = \varphi_\theta(X_i)$ at fixed θ. The derivatives of $\widehat{V}_{n,h}$ are

$$\widehat{V}_{n,h}^{(1)}(\theta) = -2n^{-1} \sum_{i=1}^n \{Y_i - \widehat{g}_{n,h}(Z_i)\} \widehat{g}_{n,h}^{(1)}(Z_i) \varphi_\theta^{(1)}(X_i)$$

$$\widehat{V}_{n,h}^{(2)}(\theta) = -2n^{-1} \sum_{i=1}^n [\{Y_i - \widehat{g}_{n,h}(Z_i)\} \widehat{g}_{n,h}^{(2)}(Z_i) - \widehat{g}_{n,h}^{(1)2}(Z_i)] \{\varphi_\theta^{(1)}(X_i)\}^{\otimes 2}$$
$$- n^{-1} \sum_{i=1}^n \{Y_i - \widehat{g}_{n,h}(Z_i)\} \widehat{g}_{n,h}^{(1)}(Z_i) \varphi_\theta^{(2)}(X_i)$$

where $-\widehat{V}_{n,h}^{(2)}(\theta)$ is a positive definite matrix converging to a finite limit $\Sigma_V^{(2)}(\theta)$ uniformly on the parameter space. The mean of $\widehat{V}_{n,h}^{(1)}(\theta)$ and its variance have the same orders as for the derivative of (7.3) in model (7.1)

$$V_{n,h}^{(1)}(\theta) = 2E[\{Y_i - \widehat{g}_{n,h} \circ \varphi_\theta(X_i)\} \widehat{g}_{n,h}^{(1)} \circ \varphi_\theta(X_i) \varphi_\theta^{(1)}(X_i)]$$
$$= 2E\{[\{(m - g_{n,h}) g_{n,h}^{(1)} - Cov(\widehat{g}_{n,h}^{(1)}, \widehat{g}_{n,h})\} \circ \varphi_\theta \varphi_\theta^{(1)}](X_i)\}$$

where the first terms in the right-hand side is $O(h^2)$ and the last term is a $O((nh^2)^{-1})$. The variance of $\widehat{V}_{n,h}^{(1)}(\theta)$ is a $O((nh^3)^{-1})$. An expansion of $V_{n,h}$ in a neighborhood of θ_0 implies

$$(nh^3)^{1/2}(\widehat{\theta}_{n,h} - \theta_0) = \{-\widehat{V}_{n,h}^{(2)}(\theta_0)\}^{-1} (nh^3)^{1/2} \{\widehat{V}_{n,h}^{(1)}(\theta_0) - V^{(1)}(\theta_0)\} + o_p(1).$$

The variance of $\widehat{V}_{n,h}^{(1)}(\theta_0)$ is asymptotically equivalent to $(nh^3)^{-1}\Sigma_{V,\theta_0} + o((nh^3)^{-1})$, with a modified notation

$$\Sigma_{V,\theta_0} = 4E[\varphi_{\theta_0}^{(1)\otimes 2}(X) f_X^{-2}(X) \{g^2 + \sigma_g^2\} \circ \varphi_{\theta_0}(X) w_2(\varphi_{\theta_0}(X))] (\int K^{(1)2}).$$

Proposition 7.3. *Under Conditions 2.1 and 3.1, and with a bandwidth $h_n = o(n^{-1/7})$ for the estimation of a regression function m in class C_2, the estimators of the parameter θ and the function m are consistent, $(nh^3)^{1/2}(\widehat{\theta}_{n,h}-\theta_0)$ converges weakly to a centered Gaussian with variance v_θ and $(nh^3)^{1/2}(\widehat{m}_{n,h}-m_{\theta_0})$ converges weakly to a centered Gaussian process with covariance $g^{(1)}\circ\varphi_{\theta_0}(x)v_{\theta_0}g^{(1)}\circ\varphi_{\theta_0}(x')\varphi^{(1)}_{\theta_0}(x)\otimes\varphi^{(1)}_{\theta_0}(x')$ at (x,x').*

With the optimal bandwidth $h_n = O(n^{-1/7})$, the convergence rates of the estimators are $n^{2/7}$ and the limiting distributions of the estimators have a non zero mean, as in the previous section.

The differential criterion $\widehat{W}_{n,h}(7.6)$ adapted to model (7.7) with a constant variance is written

$$\widehat{W}_{n,h}(\theta) = n^{-2}\sum_{i,j=1}^n \{Y_i - Y_j - \widehat{g}_{n,h}\circ\varphi_\theta(X_i) + \widehat{g}_{n,h}\circ\varphi_\theta(X_j)\}^2 K_h(X_i - X_j),$$

it defines the estimators $\widetilde{\theta}_{n,h} = \arg\min_{\theta\in\Theta}\widehat{W}_{n,h}(\theta)$ and $\widetilde{g}_{n,h}$ at $\widetilde{\theta}_{n,h}$. Let $Z_i = \varphi_\theta(X_i)$ at fixed θ, the derivatives of $\widehat{W}_{n,h}$ with respect to θ are

$$\begin{aligned}\widehat{W}^{(1)}_{n,h}(\theta) &= -2n^{-2}\sum_{i,j=1}^n \{Y_i - Y_j - \widehat{g}_{n,h}(Z_i) + \widehat{g}_{n,h}(Z_j)\}\\ &\quad \{\widehat{g}^{(1)}_{n,h}(Z_i)\varphi^{(1)}_\theta(X_i) - \widehat{g}^{(1)}_{n,h}(Z_j)\varphi^{(1)}_\theta(X_j)\}K_h(X_i - X_j),\\ \widehat{W}^{(2)}_{n,h}(\theta) &= 2n^{-2}\sum_{i,j=1}^n [\{\widehat{g}^{(1)}_{n,h}(Z_i)\varphi^{(1)}_\theta(X_i) - \widehat{g}^{(1)}_{n,h}(Z_j)\varphi^{(1)}_\theta(X_j)\}^2\\ &\quad - \{Y_i - Y_j - \widehat{g}_{n,h}(Z_i) + \widehat{g}_{n,h}(Z_j)\}\{\widehat{g}^{(2)}_{n,h}(Z_i)\varphi^{(1)2}_\theta(X_i)\\ &\quad - \widehat{g}^{(2)}_{n,h}(Z_j)\varphi^{(1)2}_\theta(X_j)\}^2 + \{Y_i - Y_j - \widehat{g}_{n,h}(Z_i) + \widehat{g}_{n,h}(Z_j)\}\\ &\quad \{\widehat{g}^{(1)}_{n,h}(Z_i)\varphi^{(2)}_\theta(X_i) - \widehat{g}^{(1)}_{n,h}(Z_j)\varphi^{(2)}_\theta(X_j)\}]K_h(X_i - X_j).\end{aligned}$$

The mean of the first derivative is $W^{(1)}_{n,h}(\theta) = -E\big(E[\{g(Z_i) - g(Z_j) - \widehat{g}_{n,h}(Z_i)+\widehat{g}_{n,h}(Z_j)\}\{\widehat{g}^{(1)}_{n,h}(Z_i)\varphi^{(1)}_\theta(X_i)-\widehat{g}^{(1)}_{n,h}(Z_j)\varphi^{(1)}_\theta(X_j)\} \mid X_i, X_j]K_h(X_i - X_j)\big)$, its order and the order of its variance are the same as in the single index model. The second derivative has the mean

$$\begin{aligned}W^{(2)}_{n,h}(\theta) = 2E\big(&E[\{\widehat{g}^{(1)}_{n,h}(Z_i)\varphi^{(1)}_\theta(X_i) - \widehat{g}^{(1)}_{n,h}(Z_j)\varphi^{(1)}_\theta(X_j)\}^2 \mid X_i, X_j]\\ &- E[\{g(Z_i) - g(Z_j) - \widehat{g}_{n,h}(Z_i) + \widehat{g}_{n,h}(Z_j)\}\{\widehat{g}^{(2)}_{n,h}(Z_i)\varphi^{(1)2}_\theta(X_i)\\ &\qquad - \widehat{g}^{(2)}_{n,h}(Z_j)\varphi^{(1)2}_\theta(X_j)\}^2 \mid X_i, X_j]\\ &+ E[\{g(Z_i) - g(Z_j) - \widehat{g}_{n,h}(Z_i) + \widehat{g}_{n,h}(Z_j)\}\{\widehat{g}^{(1)}_{n,h}(Z_i)\varphi^{(2)}_\theta(X_i)\\ &\qquad - \widehat{g}^{(1)}_{n,h}(Z_j)\varphi^{(2)}_\theta(X_j)\} \mid X_i, X_j]K_h(X_i - X_j)\big).\end{aligned}$$

The results of Proposition 7.2 are similar with these notations for the asymptotic variances.

In a regression model for processes $(X_t, Y_t)_{t\geq 0}$, empirical error processes are defined as in Section 3.6, with linear combinations of the components of the d-dimensional regression variable. They are indexed by T, the length of the time interval and the convergence rates are similar replacing n by T.

Varying bandwidth estimators of the parameter θ and the function m are defined by a modification of the estimators of the functions g and σ^2 as in Section 4.3. Assuming that the functions g and σ^2 have the same order of derivability, both functions are estimated with bandwidths of the same order, and the modified estimators are introduced in the definition of $\widehat{V}_{n,h}$, where the bandwidths $h_n(X_i)$ and $h_n(X_j)$ differ. Under Condition 4.1, Proposition 7.1 extends to variable bandwidth estimators, with the convergence rate $(n\|h_n\|^3)^{1/2}$ for the parameter estimator.

The differential empirical error uses another weight $K_h(X_i - X_j)$, for every $i \neq j$. Its bandwidth was supposed equal to the bandwidth used for the estimation of g in Section 2. Due to the symmetry of the kernel with respect to the variables X_i and X_j, a variable bandwidth in the expression $K_h(X_i - X_j)$ can be chosen as the mean of the bandwidths at X_i and X_j, with $|X_i - X_j| < 2\|h\|$. As h tends to zero, $h(X_i) = h(X_j) + (X_i - X_j)\{h^{(1)}(X_j) + o(\|h\|)\}$ and the mean bandwidth for $K_h(X_i - X_j)$ is equivalent to $h(X_i)$ or $h(X_j)$. The expansions of Chapter 4 allow to extend Proposition 7.2 for such bandwidths.

7.4 Exercises

(1) Write the mean, the bias and the variance of the local mean squared error $\widehat{V}_{n,h}(\theta; z)$ and define a sequence of optimal bandwidth functions for this criterium.

(2) Write the derivatives of $\widehat{V}_{n,h}(\theta; z)$ and an approximation for the estimator $\widehat{\theta}_{n,h}(z)$ minimizing $\widehat{V}_{n,h}(\theta; z)$. Determine the orders of its bias and its variance.

Chapter 8

Diffusion processes

8.1 Introduction

Let α and β be two functions of class C_2 on a functional metric space $(\mathbb{X}, \|\cdot\|)$, and let B be the standard Brownian motion on $\mathbb{R}$. Their norms on $\mathbb{X}$, $\|\alpha\|_1$, $\|\beta\|_2$, $E\|\alpha(X(t))\|_1$ and $E\|\beta(X(t))\|_2$ are supposed to be finite. A diffusion process $(X_t)_{t\in[0,T]}$ is defined as a stochastic differential equation by

$$dX_t = \alpha(X_t)dt + \beta(X_t)dB_t, t \in [0,T] \tag{8.1}$$

and its initial value X_0 such that $E|X_0| < \infty$. Equation (8.1) with locally Lipschitz drift and diffusion functions has a unique solution $X_t = X_0 + \int_0^t \alpha(X_s)ds + \int_0^t \beta(X_s)dB_s$, it is a continuous Gaussian process wih mean $E(X_t - X_0) = \int_0^t E\alpha(X_s)\,ds$ and variance

$$\begin{aligned} Var(X_t - X_0) &= Var\{\int_0^t \alpha(X_s)\,ds\} + E\int_0^t \beta^2(X_s)\,ds \\ &\leq \int_0^t E\{\alpha^2(X_s) + \beta^2(X_s)\}\,ds - \{\int_0^t E\alpha(X_s)\,ds\}^2. \end{aligned}$$

The existence and unicity of the process X is proved by construction of a sequence of processes satisfying Equation (8.1) and starting from X_0 and satisfying

$$X_{n,t}-X_{n-1,t} = \int_0^t \{\alpha(X_{n,s})-\alpha(X_{n-1,s})\}ds+\int_0^t \{\beta(X_{n,s})-\beta(X_{n-1,s})\}dB_t,$$

hence $X_{n,t}$ is the finite sum from X_0 to $X_{n,t}-X_{n-1,t}$, where the convergence of the sum is a consequence of the Lischitz property of α and β. By a discretization of the time interval $[0,T]$ in n sub-intervals of length tending to zero as n tends to infinity, Equation (8.1) is approximated by

$$Y_i \equiv Y_{t_{i+1}} = X_{t_{i+1}} - X_{t_i} = (t_{i+1} - t_i)\alpha(X_{t_i}) + \beta(X_{t_i})\{B_{t_{i+1}} - B_{t_i}\}, \tag{8.2}$$

considering the functions α and β as piecewise constant on the intervals of the partition generated by $(t_i)_{i=1,\dots,n}$. Let $\varepsilon_i = B_{t_{i+1}} - B_{t_i}$, it is a random variable with mean zero and variance $(t_{i+1} - t_i)$ conditionally on the σ-algebra $\mathcal{F}_{t_i}$ generated by the sample-paths of X up to t_i, then $E\alpha(X_{t_i})\varepsilon_i = 0$ and $Var(Y_i|X_{t_i}) = (t_{i+1} - t_i)\beta^2(X_{t_i})$. The process X_t solution of (8.1) is a continuous Gaussian process with independent increments. Its increments are approximated by the nonparametric regression model (8.2) with an independent normal error by considering the functions α and β as stepwise constant functions on the partition $(t_i)_{1\le i\le n}$. In the nonparametric regression model (8.2), $EY_i = (t_{i+1} - t_i)E\alpha(X_{t_i})$ and $VarY_i = (t_{i+1} - t_i)^2 Var\alpha(X_{t_i}) + (t_{i+1} - t_i)E\beta^2(X_{t_i})$.

Let t in $I_i =]t_i, t_{i+1}]$, the approximation error of the process $X_t - X_{t_i}$ by the discretized sample-path (8.2) is $e_{t;t_i} = \int_{t_i}^t \{\alpha(X_s) - \alpha(X_{t_i})\}ds + \int_{t_i}^t \{\beta(X_s) - \beta(X_{t_i})\}dB_s$, it satisfies

$$
\begin{aligned}
E|e_{t,t_i}| &\le (t - t_i)^2 \sup_{s\in I_i} |\alpha^{(1)}(X_s)|,\\
Var\, e_{t,t_i} &= Var \int_{t_i}^t \{\alpha(X_s) - \alpha(X_{t_i})\}ds + \int_{t_i}^t E\{\beta(X_s) - \beta(X_{t_i})\}^2 ds\\
&\le \int_{t_i}^t Var\{\alpha(X_s) - \alpha(X_{t_i})\}ds + \int_{t_i}^t E\{\beta(X_s) - \beta(X_{t_i})\}^2 ds
\end{aligned}
$$

and it bounded by $(t - t_i)\|\alpha^{(1)}\|^2 \,\|\beta^{(1)}\|^2\, E(X_t - X_{t_i})^2 = O((t - t_i)^2)$, with

$$
E(X_t - X_{t_i})^2 \le \int_{t_i}^t E\{\alpha^2(X_s) + \beta^2(X_s)\}\, ds.
$$

The following condition allows to express the moments of increments of the diffusion process as integrals with respect to a mean density.

Condition 8.1. There exists a mean density of the variables $(X_{t_i})_{1\le i\le n}$ defined as the limit

$$
f(x) = \lim_{n\to\infty} n^{-1} \sum_{i=1}^n f_{X_{t_i}}(x) = Ef_{X_t}(x).
$$

This condition is satisfied under a mixing property of the process X_t

$$
\begin{aligned}
&\sup\{\Pr(B|A) - \Pr(B); A \in \mathcal{F}_0^t, B \in \mathcal{F}_{t+s}^\infty, s, t \in \mathbb{R}_+\} \le \varphi(s),\\
&\int_0^T \varphi(u)\, du < \infty,
\end{aligned}
\tag{8.3}
$$

where the σ-algebras $\mathcal{F}_0^t$ and $\mathcal{F}_{t+s}^\infty$ are respectively generated by $\{X_u, u \in [0, t]\}$ and $\{X_u, u \in [t + s, \infty[\}$. That property is satisfied for the Brownian motion, its sample paths having independent increments. For the diffusion process X_t, it is sufficient that $E\int_0^T \beta^2(X_s)\,ds < \infty$. Moments of discontinuous parts of a diffusion process with jumps require another ergodicity condition defining another mean density and it is satisfied under the mixing property (8.3).

8.2 Estimation for continuous diffusions by discretization

The regression model (8.2) with observations at fixed points regularly spaced on a grid $(t_i)_{1\leq i\leq n}$, of path $\Delta_n = n^{-1}T$, is written $Y_i = \Delta_n\alpha(X_{t_i}) + \beta(X_{t_i})\varepsilon_i$, for $i = 1,\ldots,n$. The variables ε_i have a normal distribution $\mathcal{N}(0,\Delta_n)$, hence $E\varepsilon_i^{2k+1} = 0$ for every integer k, $E\varepsilon_i^2 = \Delta_n$ and $E\varepsilon_i^{2k} = (2k-1)\Delta_n E\varepsilon_i^{2(k-1)}$ for every $k \geq 1$, thus $E\varepsilon_i^4 = 3\Delta_n^2$. A nonparametric estimator of the function α requires a normalization of Y_i by the scale Δ_n^{-1}

$$\widehat{\alpha}_{n,h}(x) = \frac{\sum_{i=1}^n Y_i K_h(x - X_{t_i})}{\Delta_n \sum_{i=1}^n K_h(x - X_{t_i})},$$

for every x in $\mathcal{X}_{n,h} = \{y \in \mathcal{X}; \|x-y\| < h\}$. The approximations and the convergences of Proposition 3.1 are satisfied for the estimator $\widehat{\alpha}_{n,h}$. Let $\alpha_{n,h}$ be its mean, it is approximated as $\alpha_{n,h}(x) = \mu_{\alpha,n,h}(x) f_{X,n,h}^{-1}(x) + O((Th)^{-1})$, where

$$\mu_{\alpha,n,h}(x) = \Delta_n^{-1} \int\int y K_h(x-s)\,dF_{X_t,Y_t}(s,y)$$
$$\mu_\alpha(x) = E f_{X_t}(x)\alpha(x).$$

The bias of the estimator $\widehat{\alpha}_{n,h}$ for a function α in class $C_2(\mathcal{X})$ has a first order expansion

$$b_{\alpha,n,h}(x) = \alpha_{n,h}(x) - \alpha(x) = h^2 b_\alpha(x) + o(h^2),$$
$$b_\alpha(x) = \frac{m_{2K}}{2} f^{-1}(x)\{\mu_\alpha^{(2)}(x) - \alpha(x) f^{(2)}(x)\}$$

and its variance is

$$v_{\alpha,n,h}(x) = \Delta_n^{-1}(nh)^{-1}\{\sigma_\alpha^2(x) + o(1)\},$$
$$\sigma_\alpha^2(x) = \kappa_2 f^{-1}(x) Var(Y_t \mid X_t = x) = \kappa_2 f^{-1}(x)\beta^2(x).$$

For the estimation of the function β defining the variance of X, let

$$Z_i = Y_i - \Delta_n\widehat{\alpha}_{n,h}(X_{t_i}) = \Delta_n(\alpha - \widehat{\alpha}_{n,h})(X_{t_i}) + \beta(X_{t_i})\varepsilon_i,$$

its mean is $E\{Z_i|X_{t_i}=x\}=\Delta_n E(\alpha-\widehat{\alpha}_{n,h})(X_{t_i})=\Delta_n(\alpha-\alpha_{n,h})(X_{t_i})$ its order is $\Delta_n O(h^2)$ and its variance satisfies

$$\begin{aligned}
\Delta_n^{-1}Var\{Z_i|X_{t_i}=x\} &= \Delta_n^{-1}E\{\Delta_n(\alpha_{n,h}-\widehat{\alpha}_{n,h})(X_{t_i})+\beta(X_{t_i})\varepsilon_i\}^2 \\
&= \Delta_n Var\widehat{\alpha}_{n,h}(X_{t_i})+\beta^2(x) \\
&= (nh)^{-1}\kappa_2 f^{-1}(x)+\beta^2(x). \qquad (8.4)
\end{aligned}$$

A consistent estimator of the function $\beta^2(x)$ is therefore

$$\widehat{\beta}^2_{n,h}(x)=\frac{\sum_{i=1}^n Z_i^2 K_h(x-X_{t_i})}{\Delta_n\sum_{i=1}^n K_h(x-X_{t_i})}.$$

The approximations and the convergences of Proposition 3.1 are also satisfied for the estimator $\widehat{\beta}_n$. Let $\beta_{n,h}$ be its mean and let

$$\begin{aligned}
\mu_{\beta,n,h}(x) &= \Delta_n^{-1}E[Z_i^2K_h(x-X_{t_i})] \\
&= E\big([\beta^2(X_t)+(nh)^{-1}\kappa_2 f^{-1}(X_{t_i}) \\
&\qquad -\Delta_n(\alpha_{n,h}-\alpha)^2(X_t)\}]K_h(x-X_t)\big) \\
&= \mu_\beta(x)+o(1), \\
\mu_\beta(x) &= f(x)\beta^2(x).
\end{aligned}$$

The mean $\beta_{n,h}$ is approximated as $\beta^2_{n,h}(x) = \mu_{\beta,n,h}(x)f^{-1}_{X,n,h}(x)+O((nh)^{-1})$. Under conditions (2.1) and (3.1) for the functions α and β in class $C_2(\mathcal{X})$, the bias of the estimator $\widehat{\beta}_n$ is

$$\begin{aligned}
b_{\beta^2,n,h}(x) &= \beta^2_{n,h}(x)-\beta^2(x)=h^2 b_\beta(x)+o(h^2), \\
b_{\beta^2}(x) &= \frac{m_{2K}}{2}f^{-1}(x)\{\mu_\beta^{(2)}(x)-\beta^2(x)f^{(2)}(x)\}
\end{aligned}$$

and its variance is $v_{\beta,n,h}(x)=(nh)^{-1}\{\sigma^2_\beta(x)+o(1)\}$ where $\sigma^2_\beta(x)$ is the first term in the expansion of $\kappa_2 f^{-1}(x)\Delta_n^{-2}Var(Z_t^2\mid X_t=x)$, provided by the approximation of $\Delta_n^{-2}Var(Z_t^2\mid X_t=x)$ by

$$\begin{aligned}
&E\{\Delta_n^2(\widehat{\alpha}_{n,h}-\alpha)^4(X_t)+\beta^4(X_t)\Delta_n^{-2}\varepsilon^4 \\
&\qquad +2\beta^2(X_t)(\widehat{\alpha}_{n,h}-\alpha)^2(X_t)\mid X_t=x\}-E^2(\Delta_n^{-1}Z_t^2\mid X_t=x) \\
&= \beta^4(x)\Delta_n^{-2}E\varepsilon^4+\Delta_n^2E(\widehat{\alpha}_{n,h}-\alpha)^4(x)+2\beta^2(x)\{v_{\alpha,n,h}(x)+b^2_{\alpha,n,h}(x)\} \\
&\qquad -\{\beta^2(x)+v_{\alpha,n,h}(x)+b^2_{\alpha,n,h}(x)\}^2 \\
&= \beta^4(x)\{\Delta_n^{-2}E\varepsilon^4-1\}=2\beta^4(x)+o(1),
\end{aligned}$$

thus σ^2_β is written in the form $\sigma^2_\beta(x)=2\kappa_2 f^{-1}(x)\beta^4(x)$. Under the condition $h=h_T=0(T^{-1/5})$, let $c_\alpha=\lim_{T\to\infty}(Th_T^5)^{1/2}$.

Proposition 8.1. *Under Conditions 2.1, 2.2, 3.1 and 3.2 for the functions α and β in class $C_s(\mathcal{X})$, the estimators $\widehat{\alpha}_{n,h}$ and $\widehat{\beta}_{n,h}$ are uniformly consistent on $\mathcal{X}$, with bias*

$$\begin{aligned} b_{\alpha,n,h}(x) &= \alpha_{n,h}(x) - \alpha(x) = h^s b_\alpha(x) + o(h^s), \\ b_\alpha(x) &= \frac{m_{sK}}{s!} f^{-1}(x)\{\mu_\alpha^{(s)}(x) - \alpha(x) f^{(s)}(x)\}, \\ b_{\beta^2,n,h}(x) &= \beta^2_{n,h}(x) - \beta^2(x) = h^s b_\beta(x) + o(h^s), \\ b_{\beta^2}(x) &= \frac{m_{sK}}{s!} f^{-1}(x)\{\mu_\beta^{(s)}(x) - \beta^2(x) f^{(s)}(x)\} \end{aligned}$$

and their variances are $v_{\alpha,n,h}(x)$ and $v_{\beta,n,h}(x)$. Moreover

$$\|\widehat{\alpha}_{n,h}(x) - \alpha_{n,h}(x)\|_p = 0((Th)^{-1/p})\ \|\widehat{\beta}_{n,h}(x) - \beta_{n,h}(x)\|_p = 0((nh)^{-1/p}),$$

for every $p \geq 2$, where the approximations are uniform. If $h = 0(T^{-1/5})$, the process $(Th)^{1/2}(\widehat{\alpha}_{n,h} - \alpha - c_\alpha b_\alpha)$ converges weakly to centered Gaussian process with variance $\sigma^2_\alpha(x)$, the process $(nh)^{1/2}(\widehat{\beta}^2_{n,h} - \beta^2 - \gamma^{1/2} b_{\beta^2})$ converges weakly to centered Gaussian process with variance $\sigma^2_\beta(x)$ at x, and the limiting covariances are zero.

The order for the bandwidths is the order of the optimal bandwidth for the asymptotic mean squared errors of estimation of α. The conditions ensure a Lipschitz property for the second order moment of the increments of the processes, similar to Lemma 2.2 for the density. Moreover, the covariances develop like in the proof of Theorem 2.1.

The variance of the variable Y in model (8.2) being a function of X, the regression function α is also estimated by the mean of a weighted kernel as in Section 3.6, with the weighting variables $\widehat{w}(X_{t_i}) = \sigma_\alpha^{-1}(X_{t_i})$. As previously, the approximations of the bias and variance of the new estimator (3.19) of the drift function are modified by introducing $\widehat{w}_n$ and its asymptotic distribution is modified.

With a partition of $[0,T]$ in subintervals I_i of unequal length $\Delta_{n,i}$ varying with the observation timest$_i$ of the process the variable Y_i has to be normalized by $\Delta_{n,i}$, $1 \leq i \leq n$. For every x in $\mathcal{X}_{n,h}$, the estimators are

$$\begin{aligned} \widehat{\alpha}_{n,h}(x) &= \frac{\sum_{i=1}^n \Delta_{n,i}^{-1} Y_i K_h(x - X_{t_i})}{\sum_{i=1}^n K_h(x - X_{t_i})}, \\ Z_{n,i} &= \Delta_{n,i}^{-1/2}\{Y_i - \Delta_{n,i}\widehat{\alpha}_{n,h}(X_{t_i})\}, \\ \widehat{\beta}^2_{n,h}(x) &= \frac{\sum_{i=1}^n Z^2_{n,i} K_h(x - X_{t_i})}{\sum_{i=1}^n K_h(x - X_{t_i})}. \end{aligned}$$

The results of Proposition 8.1 are satisfied, replacing the means of sums with terms $\Delta_{n,i}^{-1}$ by means with coefficient $n^{-1}\sum_{i=1}^{n}\Delta_{n,i}^{-1}$ and assuming that the lengths $\Delta_{n,i}$ have the order $n^{-1}T$. The optimal bandwidth for the estimation of α is $O(T^{-1/(2s+1)})$ and its asymptotic mean squared error is $AMSE_\alpha(x) = (Th)^{-1}\sigma_\alpha^2(x) + h^{s2}b_{\alpha,s}^2(x)$, it is minimum for the bandwidth function

$$h_{\alpha,AMSE}(x) = \left\{\frac{(s!)^2\kappa_2}{2sm_{sK}^2}\frac{T^{-1}Var(\Delta_{n,i}^{-1}Y_{t_i})}{\{\mu_\alpha^{(s)}(x) - \alpha(x)f^{(s)}(x)\}^2}\right\}^{1/(2s+1)}.$$

The optimal local bandwidth for estimating the variance function β^2 of the diffusion is a $O(n^{-1/(2s+1)})$ and it minimizes $AMSE_\beta(x) = (nh)^{-1}\sigma_\beta^2(x) + h^{s2}b_{\beta,s}^2(x)$.

A diffusion model including several explanatory processes in the coefficients α and β may be written using an indicator process $(J_t)_t$ with values in a discrete space $\{1,\ldots,K\}$ as

$$dX_t = \sum_{k=1}^{K}\alpha_k(X_t)1\{J_t = k\}dt + \sum_{k=1}^{K}\beta_k(X_t)1\{J_t = k\}dB_t, t \in [0,T]. \quad (8.5)$$

Let $X_{tk} = X_t1\{J_t = k\}$ be the partition of the variable corresponding to the models for the drift and the variance of equation (8.5). The model is equivalent to

$$dX_t = \sum_{k=1}^{K}\alpha_k(X_{tk})dt + \sum_{k=1}^{K}\beta_k(X_{tk})dB_t$$

and the estimators of the $2K$ functions α_k and β_k are defined for every for x in $\mathcal{X}_{n,h}$ by

$$\widehat{\alpha}_{k,n,h}(x) = \frac{\sum_{i=1}^{n}\Delta_{n,i}^{-1}Y_iK_h(x - X_{t_i,k})}{\sum_{i=1}^{n}K_h(x - X_{t_i,k})},$$
$$\widehat{\beta}_{k,n,h}^2(x) = \frac{\sum_{i=1}^{n}\Delta_{n,i}^{-1}Z_i^2K_h(x - X_{t_i,k})}{\sum_{i=1}^{n}K_h(x - X_{t_i,k})}.$$

Their means are approximated as the estimators for model (8.1) by

$$\alpha_{k,n,h}(x) = \mu_{\alpha_k,n,h}(x)f_{X,n,h}^{-1}(x) + O((Th)^{-1}),$$
$$\beta_{k,n,h}^2(x) = \mu_{\beta_k,n,h}(x)f_{X,n,h}^{-1}(x) + O((nh)^{-1}),$$

where

$$\begin{aligned}
\mu_{\alpha,k,n,h}(x) &= En^{-1}\sum_{i=1}^{n}\Delta_{n,i}^{-1}\int\int yK_h(x-s)\,dF_{X_{k,t_i},Y_{t_i}}(s,y) \\
&= \mu_{\alpha_k}(x) + \frac{h^2}{2}m_{2K}\mu_{\alpha_k}^{(2)}(x) + o(h^2) \\
\mu_{\alpha_k}(x) &= f(x)\alpha_k(x), \\
\mu_{\beta_k,n,h}(x) &= n^{-1}\sum_{i=1}^{n}\Delta_{n,i}^{-1}E[Z_i^2K_h(x-X_{k,t_i})] \\
&= \mu_{\beta_k}(x) + \frac{h^2}{2}m_{2K}\mu_{\beta_k}^{(2)}(x) + o(h^2), \\
\mu_{\beta_k}(x) &= f(x)\beta_k^2(x).
\end{aligned}$$

The norms and the asymptotic behaviour of the estimators is the same as in Proposition 8.1. The two-dimensional model

$$\begin{aligned}
dX_t &= \alpha_X(Y_t)dt + \beta_X(X_t)dB_X(t), \\
dY_t &= \alpha_Y(Y_t)dt + \beta_Y(Y_t)dB_Y(t)
\end{aligned}$$

with independent Brownian processes B_X and B_Y is a special case where all parameters are estimated as before.

The process (8.1) is generalized with functions depending of the sample-path and of the current time

$$dX_t = \alpha(t, X_t)dt + \beta(t, X_t)dB_t, t \in [0, T] \tag{8.6}$$

under similar conditions. A discretization of the time interval $[0, T]$ leads to

$$Y_i = X_{t_{i+1}} - X_{t_i} = (t_{i+1} - t_i)\alpha(t_i, X_{t_i}) + \beta(t_i, X_{t_i})(B_{t_{i+1}} - B_{t_i}).$$

The functions α and β are now defined in $(\mathbb{R}_+ \times \mathbb{X})$ and they are estimated by

$$\begin{aligned}
\widehat{\alpha}_{n,h}(t,x) &= \frac{\sum_{i=1}^{n} Y_i K_{h_1}(x-X_{t_i})K_{h_2}(t-t_i)}{\Delta_n\sum_{i=1}^{n} K_{h_1}(x-X_{t_i})K_{h_2}(t-t_i)}, \\
\widehat{\beta}_{n,h}^2(t,x) &= \frac{\sum_{i=1}^{n} Z_i^2 K_{h_1}(x-X_{t_i})K_{h_2}(t-t_i)}{\Delta_n\sum_{i=1}^{n} K_{h_1}(x-X_{t_i})K_{h_2}(t-t_i)},
\end{aligned}$$

with $Z_i = Y_i - \Delta_n\widehat{\alpha}_{n,h}(t_i, X_{t_i}) = \Delta_n(\alpha - \widehat{\alpha}_{n,h})(t_i, X_{t_i}) + \beta(t_i, X_{t_i})\varepsilon_i$. Their variance has now the order $(nh_1h_2)^{-1}$, their bias is a $O((h_1h_2)^2)$ and the convergence rate of the centered processes is $(nh_1h_2)^{1/2}$.

8.3 Estimation for continuous diffusion processes

The process $\{X_t, t \in [0,1]\}$ is extended to a time interval $[0,T]$ by rescaling: $X_t = X_{Ts}$, with s in $[0,1]$ and t in $[0,T]$. Now the Gaussian process B is mapped from $[0,1]$ onto $[0,T]$ by the same transform and $B_s = T^{1/2}B_{t/T}$ is the Brownian motion extended from $[0,1]$ to $[0,T]$. The observation of the sample-path of the process $\{X_t, t \in [0,T]\}$ allows to construct estimators similar those of smooth density and regression function in Sections 2.10 and 3.10, under the ergodic property (2.13). The Brownian process $(B_t)_{t\geq 0}$ is a martingale with respect to the filtration generated by the $(B_u)_{u<t}$, $E(B_t - B_s \mid X_s) = 0$ for every $0 < s < t$. Its moments are $EB_t^{2k+1} = 0$, $B_t^2 = t$ thus $(B_t - B_0)^2$ has a $t\chi_1^2$ distribution and, for every integer k, $B_t^{2k} = t^k G^{(k)}(0)$ with the generating function $G_{2k}(t) = (1-2t)^{-k}$ of the χ_k^2 distribution, for $t < 1/2$, hence $B_t^4 = 3t^2$.

Estimators are built like for regression functions of processes with the response process $Y_t = dX_t$, without derivability assumption for the sample-paths of X since B has only a L_2-derivative. The integrated drift function

$$A(t;X) = \int_0^t \alpha(X_s)\,ds$$

is estimated by $\widehat{A}(t;X) = X_t - X_0$, thus $E\widehat{A}(t;X) = A(t;X)$ and its variance equals $Var\{\int_0^t \alpha(X_s)\,ds\} + E\int_0^t \beta^2(X_s)\,ds$. The drift function $\alpha(X_t)$ is estimated by smoothing the sample-path of the process X in a neighborhood of $X_t = x$

$$\widehat{\alpha}_{T,h}(x) = \frac{\int_0^T K_h(x - X_s)\,dX_s}{\int_0^T K_h(x - X_s)\,ds}. \tag{8.7}$$

The estimators of the density and $\mu_\alpha(x) = \alpha(x)f_X(x)$ defining (8.7) are

$$\widehat{f}_{X,T,h}(X_t) = T^{-1}\int_0^T K_h(x - X_s)\,ds,$$

$$\widehat{\mu}_{\alpha,T,h}(x) = T^{-1}\int_0^T K_h(x - X_s)\,dX_s.$$

Their limits are expressed with the mean marginal density of the process

$$f_X(x) = \lim_{T\to\infty} T^{-1}E\int_0^T f_{X_s}(x)\,ds$$

and the mixing property of the sample path of the process X implies that

$$f_X(x) - \lim_{T\to\infty} T^{-1}E\int_0^T f_{X_s}(x)\,ds = O(T^{-1/2}).$$

Assuming that the kernel satisfies Conditions 2.1-2.2, their moments are approximated using Taylor expansions and the properties of the Brownian motion, with covariance function $E(B_s B_t) = s \wedge t$. With a diffusion process X, their expectations are

$$\begin{aligned}
E\widehat{f}_{X,T,h}(x) &= \int_{\mathcal{I}_X} K_h(u-x)\, f_X(u)\, du \\
&= f_X(x) + \frac{h^2}{2} m_{2K} f_X^{(2)}(x) + o(h^2), \\
E\widehat{\mu}_{\alpha,T,h}(x) &= T^{-1} E \int_0^T K_h(x - X_s)\{\alpha(X_s)\, ds + \beta(X_s)\, dB_s\} \\
&= \int_{\mathcal{I}_X} \alpha(u) K_h(u-x) f_X(u)\, du \\
&= \alpha(x) f_X(x) + \frac{h^2}{2} m_{2K} (\alpha f_X)^{(2)}(x) + o(h^2),
\end{aligned}$$

so the bias of the estimator of $\mu_\alpha(x)$ is $h^2 b_{\mu_\alpha}(x) = h^2 m_{2K} (\alpha f_X)^{(2)}(x)/2 + o(h^2)$. Its variance $T^{-2} E \int_0^T K_h(X_t - x)\, \{dX_t - \alpha(X_t)\, dt\}^2$ is expanded using the ergodicity property (2.16) as in Section 2.10, now the covariance of $T^{-1} \int_0^T K_h(X_s - x)\beta(X_s)\, dB_s$ and $T^{-1} \int_0^T K_h(X_t - x)\beta(X_t)\, dB_t$ as a sum $I_d(T) + I_o(T)$, where $I_d(T) = T^{-2} E \int_0^T K_h^2(X_t - x)\beta^2(X_t)\, dt$ develops as

$$\begin{aligned}
I_d(T) &= T^{-1} \int_{\mathcal{I}_X} K_h^2(u-x)\beta^2(u) f_X(u)\, du \\
&= (Th_T)^{-1} \kappa_2 \beta^2(x) f_X(x) + o((Th_T)^{-1})
\end{aligned}$$

and the expectation $I_o(T)$ is expanded using the ergodicity property, with $\int_0^T \int_0^T d(s \wedge t) = 2 \int_0^T (T-s)\, ds = T^2$ and the notation $\alpha_h(u,v) = |u-v|/2h_T$

$$\begin{aligned}
I_o(T) &= \int_{\mathcal{I}_X^2 \setminus \mathcal{D}_X} K_h(u-x) K_h(v-x) x \beta(u)\beta(v)\, dF_{X_s,X_t}(u,v)\, du\, dv \\
&= \int_{\mathcal{I}_{X \setminus \{u\}}} \int_{-1}^1 K(z - \alpha_h(u,v)) K(z + \alpha_h(u,v))\, dz \\
&\qquad \beta(u)\beta(v)\, d\pi_u(v)\, dF_X(u)\}\{1 + o(1)\}.
\end{aligned}$$

For every fixed $u \neq v$, $\alpha_{h_T}(u,v)$ tends to infinity as h_T tends to zero, then the integral $\int_{-1/2}^{1/2} K(z - \alpha_h(u,v)) K(z + \alpha_h(u,v))\, dz$ tends to zero. If $\alpha_h(u,v) = O(h_T)$, this integral does not disappear but $\pi_u(v)$ tends to zero, therefore the integral $I_o(T)$ is a $o((Th_T)^{-1})$ as T tends to infinity. Under the ergodicity condition (2.16) for sets of k finite dimensional distributions of X, the L_p-norm of the centered estimator of μ_α also satisfies $\|\widehat{\mu}_{\alpha,T,h}(x) -$

$\mu_{\alpha,T,h}(x)\|_p = O((Th_T)^{-1/p})$ and the approximation (3.2) is also satisfied for the estimator $\widehat{\alpha}_{T,h}$. It follows that the estimator $\widehat{\alpha}_{T,h}(x)$ of a drift function α in class C_s, for $s \geq 2$, has a bias and a variance

$$\begin{aligned} b_{\alpha,T,h}(x;s) &= h_T^s b_\alpha(x;s) + o(h_T^s), \\ b_\alpha(x;s) &= \frac{m_{sK}}{s!} f_X^{-1}(x)\{(\alpha f_X)^{(s)}(x) - \alpha(x) f_X^{(s)}(x)\}, \\ v_{m,T,h}(x) &= (Th_T)^{-1}\{\sigma_\alpha^2(x) + o(1)\}, \\ \sigma_\alpha^2(x) &= \kappa_2 f_X^{-1}(x)\beta^2(x) \end{aligned}$$

so they have the same expressions as in the discretized regression model (8.2), the covariance of $\widehat{\alpha}_{T,h}(x)$ and $\widehat{\alpha}_{T,h}(y)$ tends to zero. Let

$$Z_t = X_t - X_0 - \int_0^t \widehat{\alpha}_{T,h}(X_s)\,ds = \int_0^t (\alpha - \widehat{\alpha}_{T,h})(X_s)\,ds + \int_0^t \beta(X_t)\,dB_t, \tag{8.8}$$

its expectation conditionally on the filtration generated by the process X up to t^- is $E(Z_t \mid \mathcal{F}_t) = -\int_0^t b_{\alpha,T,h}(X_s)\,ds = O(h^2)$ for every $t > 0$ and the main term of its conditional variance

$$\begin{aligned} Var(Z_t \mid X_t) = Var\{&\int_0^t \widehat{\alpha}_{T,h}(X_s)ds\} + \int_0^t \beta^2(X_s)ds \\ &- 2Cov\{\int_0^t (\widehat{\alpha}_{T,h})(X_s)\,ds, \int_0^t \beta(X_s)\,dB_s\} + O((Th_T)^{-1}) \end{aligned}$$

is $\int_0^t \beta^2(X_s)ds$. The variance function $\beta^2(X_t)$ is therefore consistently estimated by

$$\widehat{\beta}_{T,h}^2(x) = \frac{2\int_0^T Z_s K_h(X_s - x)\,dZ_s}{\int_0^T K_h(X_s - x)\,ds}. \tag{8.9}$$

Under conditions (2.1) and (3.1) for the functions α and β in class $C_s(\mathcal{X})$, the bias of the estimator $\widehat{\beta}_{T,h}^2$ is

$$\begin{aligned} b_{\beta,T,h}(x) &= h^s b_\beta(x;s) + o(h^s), \\ b_\beta(x;s) &= \frac{m_{sK}}{s!} f^{-1}(x)\{(f\beta^2)^{(s)}(x) - \beta^2(x) f^{(s)}(x)\}. \end{aligned}$$

Its variance is $v_{\beta,T,h}(x) = (Th)^{-1}\{\sigma_\beta^2(x) + o(1)\}$ where $\sigma_\beta^2(x)$ is the first term in the expansion of $\kappa_2 f^{-1}(x) Var(Z_t^2 \mid X_t = x)$ calculated like in the discrete model, that is $\sigma_\beta^2(x) = 2\kappa_2 f^{-1}(x)\beta^4(x)$, as in Proposition 8.1. Under the previous conditions, the processes $(Th_T)^{1/2}(\widehat{\alpha}_{T,h} - \alpha - b_{\alpha,T,h})$ and $(Th_T)^{1/2}(\widehat{\beta}_{T,h}^2 - \beta^2 - b_{\beta^2,T,h})$ converge weakly to a centered Gaussian processes with mean zero, covariances zero and respective variance functions σ_α^2 and σ_β^2.

The mean squared error of the estimator at x for a marginal density in C_s is then

$$MISE_{T,h_T}(x) = (Th_T)^{-1}\kappa_2 f_X^{-1}(x)Var(Y \mid X = x) + h_T^{2s} b_\alpha(x;s) + o((Th_T)^{-1}) + o(h_T^{2s})$$

and the optimal local and global bandwidths minimizing the mean squared (integrated) errors are $O(T^{1/(2s+1)})$

$$h_{AMSE,T}(x) = \Big\{\frac{1}{T}\frac{\sigma_\alpha^2(x)}{2sb_\alpha^2(x;s)}\Big\}^{1/(2s+1)}$$

and, for the asymptotic mean integrated squared error criterion

$$h_{AMISE,T} = \Big\{\frac{1}{T}\frac{\int \sigma_\alpha^2(x)\,dx}{2s\int b_\alpha^2(x;s)\,dx}\Big\}^{1/(2s+1)}.$$

With the optimal bandwidth rate, the asymptotic mean (integrated) squared errors are $O(T^{2s/(2s+1)})$.

The same expansions as for the variance of $\widehat{\mu}_{T,h}(x)$ and $\widehat{f}_{X,T,h}(x)$ in Section 2.10 prove that the finite dimension distributions of the process $(Th_T)^{1/2}(\widehat{\alpha}_{T,h} - \alpha - b_{\alpha,T,h})$ and $(Th_T)^{1/2}(\widehat{\beta}_{T,h} - \beta - b_{\beta,T,h})$ converge to those of a centered Gaussian process with mean zero, covariances zero and variance functions σ_α^2 and σ_β^2. Lemma 3.3 generalizes and the increments $E\{\widehat{\alpha}_{T,h}(x) - \widehat{\alpha}_{T,h}(y)\}^2$ and $E\{\widehat{\beta}_{T,h}(x) - \widehat{\beta}_{T,h}(y)\}^2$ are approximated by $O(|x-y|^2(Th_T^3)^{-1})$ for every x and y in $\mathcal{I}_{X,h}$ such that $|x-y| \le 2h_T$. Then the processes $(Th_T)^{1/2}\{\widehat{\alpha}_{T,h} - \alpha\}\mathcal{I}_{\{\mathcal{I}_{X,T}\}}$ and $(Th_T)^{1/2}\{\widehat{\beta}_{T,h} - \beta\}\mathcal{I}_{\{\mathcal{I}_{X,T}\}}$ converge weakly to $\sigma_\alpha W_1 + \gamma^{1/2} b_\alpha$ and $\sigma_\beta W_2 + \gamma^{1/2} b_\beta$, respectively, where W_1 and W_2 are centered Gaussian processes on $\mathcal{I}_X$ with variance 1 and covariances zero. The covariance $C_{\alpha,\beta,T,h}(x,y)$ of $\widehat{\alpha}_{T,h}(x)$ and $\widehat{\beta}_{T,h}(y)$, with $2|x-y| > h_T$ develops using the approximation (3.2) as $\{f(x)f(y)T\}^{-2}[E\{\int_0^T K_h(X_s - x)\beta(X_s)\,dB_s\}\{\int_0^T K_h(X_s - y)(2Z_t\,dZ_t - \beta^2(X_t)\,dt)\} - E\alpha(x)\{\int_0^T K_h(X_s - y)\beta(X_s)\,dB_s\}(\widehat{f}_{T,h} - f_{T,h})(x) - E\beta(y)\{\int_0^T K_h(X_s - x)(2Z_t\,dZ_t - \beta^2(X_t)\,dt)\}(\widehat{f}_{T,h} - f_{T,h})(y) + \alpha(x)\beta(y)(Th_T)^{-1}Cov(\widehat{f}_{T,h}(x), \widehat{f}_{T,h}(y))$, it is therefore a $o((Th_T)^{-1})$.

According to the local optimal bandwidths defined in the previous sections, the estimators $\widehat{\alpha}_{n,h}$ and $\widehat{\beta}_{n,h}$ are calculated with a functional bandwidth sequences $(h_n(x))_n$ or $(h_T(x))_T$. The assumptions for the convergence of these sequences are similar to the assumptions for the nonparametric regression with a functional bandwidth and the results of Chapter 4 apply immediatly for the estimators of the discretized or continuous processes (8.2) and (8.1).

8.4 Estimation of discretely observed diffusions with jumps

Let α, β and γ be functions of class C_2 on a metric space $(\mathbb{X}, \|\cdot\|)$, let B be the standard Brownian motion, $M = N - \widetilde{N}$ be a centered martingale associated to a point process N, with predictable compensator $\widetilde{N}(t) = \int_0^t Y\, d\Lambda$, and such that M is independent of B. The process Y is predictable and there exists a function g defined on $[0,1]$ such that $\sup_{s\in[0,1]} |T^{-1}Y_{Ts} - g(s)|$ converges to zero in probability, the function g and the hazard function λ is supposed to be in class $C_2(\mathbb{R})$; $ET^{-1}\int_0^T \gamma^2 d\widetilde{N}$ are finite for every stopping time T.

The process X_t solution of the stochastic differential equation

$$dX_t = \alpha(X_t)dt + \beta(X_t)dB_t + \gamma(X_t)dM_t, t \in [0,T], \tag{8.10}$$

has a discrete and a continuous part. A discretization of this equation into n sub-intervals of length $\Delta_{n,i}$ tending to zero as n tends to infinity gives the approximated equation

$$Y_i = X_{t_{i+1}} - X_{t_i} = \Delta_{n,i}\alpha(X_{t_i}) + \beta(X_{t_i})\Delta B_{t_i} + \gamma(X_{t_i})\Delta M_{t_i}.$$

Let $\varepsilon_i = \Delta B_{t_i} = B_{t_{i+1}} - B_{t_i}$, with zero mean and variance $\Delta_{n,i}$ conditionally on the σ-algebra $\mathcal{F}_{t_i}$ generated by the sample-paths of X up to t_i, $\eta_i = \eta(t_{i+1})$ defined by $M_{t_{i+1}} - M_{t_i}$, with expectation zero and variance $\widetilde{N}_{t_{i+1}} - \widetilde{N}_{t_i} = O(\Delta_{n,i})$ conditionally on the σ-algebra $\mathcal{F}_{t_i}$ generated by the sample-paths of X; $E\{\alpha(X_{t_i})\varepsilon_i\} = 0$, $E\{\beta(X_{t_i})\eta_i\} = 0$, and the martingales $(B_t)_{t\geq 0}$ and $(M_t)_{t\geq 0}$ have independent increments, by definition. The functionals of the martingale M and the process $\widetilde{N}$ are estimated from the observation of the point process N, as in Chapter 4. The variables X_{T_i} are supposed to satisfy an ergodic property for the random stopping times of the counting process N, in addition to Conditions 6.2 and 8.1.

Condition 8.2. There exists a mean density of the variables X_{T_i} defined as the limit

$$f_N(x) = \lim_{T\to\infty} T^{-1} \int_0^T f_{X_s}(x)\, dN(s).$$

This condition is satisfied if the jump part of the process X_t satisfies the property (8.3) and the limit is $f_N(x) = T^{-1}E\int_0^T f_{X_s}(x)\, d\widetilde{N}(s)$. The diffusion process X_t defined by (8.10) has the mean

$$\begin{aligned}
\mu_T &= EX_0 + E\int_0^t \alpha(X_s)ds \\
&= EX_0 + \int_{\mathbb{X}}\int_0^t \alpha(x) f_{X_s}(x)\, dt\, dx = EX_0 + t\int_{\mathbb{X}} \alpha(x) f(x)\, dx
\end{aligned}$$

and the variance of the normalized variable $T^{-1/2}(X_T - \mu_T)$ is finite if the integrals

$$S_\alpha = ET^{-1}\int_0^T \alpha^2(X_s)\,ds = \int_{\mathbb{X}} \alpha^2(x)f(x)\,dx + o(1),$$
$$S_\beta = ET^{-1}\int_0^T \beta^2(X_s)\,ds = \int_{\mathbb{X}} \beta^2(x)f(x)\,dx + o(1),$$
$$S_\gamma = ET^{-1}\int_0^T \gamma^2(X_s)\,d\widetilde{N}(s) = \int_{\mathbb{X}} \gamma^2(x)f_N(x)\,dx + o(1)$$

are finite. Then $T^{1/2}(T^{-1}X_T - \mu_T)$ converges weakly to a centered Gaussian variable with variance $S_X = S_\alpha + S_\beta + S_\gamma$. Let $S_X(t)$ be the function defined as above with integrals on $[0,t]$. The process $T^{1/2}(T^{-1}X_{sT} - \mu_{sT})_{0\le s\le 1}$ is a sum of stochastic integrals with respect to the martingales B and M.

Proposition 8.2. *The process $W_{T,s} = T^{1/2}(T^{-1}X_{sT} - \mu_{sT})_{0\le s\le 1}$ is a martingale. If $S_X < \infty$, $W_{T,s}$ converges weakly to a Brownian motion B_X with variance function $S_X(s)$ on $[0,1]$.*

Let a in $\mathbb{R}$ and $T_a = \inf\{s \in [0,1]; B_X(s) = a\}$ be a stopping time for the process B_X, then for every $\theta \ge 0$

$$E\exp\{\theta S_X(T_a)\} = \exp(-a\sqrt{2\theta}).$$

Let a in $\mathbb{R}$ and $T_{T,a} = \inf\{s \in [0,1]; W_{T,s} = a\}$ be a stopping time for the process $W_{T,s}$.

Corollary 8.1. *For every $\theta \ge 0$, $E\exp\{\theta S_X(T_{T,a})\}$ converges to $\exp(-a\sqrt{2\theta})$ as T tends to infinity.*

Moments of discontinuous parts of a diffusion process with jumps require another ergodicity condition defining another mean density and it is satisfied under the mixing property (8.3). Conditionally on $\mathcal{F}_{t_i}$, the variables Y_i have the expectation $\Delta_{n,i}\alpha(X_{t_i})$ and the variance

$$\begin{aligned} Var(Y_i|X_{t_i}) &= \beta^2(X_{t_i})\Delta_{n,i} + \int_{t_i}^{t_{i+1}} \gamma^2(X_s)\,d\widetilde{N}(s) \\ &= \beta^2(X_{t_i})\Delta_{n,i} + \gamma^2(X_{t_i})\,\Delta\widetilde{N}(t_i) + o(\Delta_{n,i}) = O(\Delta_{n,i}). \end{aligned}$$

A nonparametric estimator of the function α is the kernel estimator normalized by $\Delta_{n,i}$ as in the previous section

$$\widehat{\alpha}_{n,h}(x) = \frac{\sum_{i=1}^n \Delta_{n,i}^{-1} Y_i K_h(x - X_{t_i})}{\sum_{i=1}^n K_h(x - X_{t_i})},$$

for x in $\mathcal{X}_{n,h}$. Let Δ_n^{-1} denote $n^{-1}\sum_{i=1}^n \Delta_{n,i}^{-1}$ and $E\widetilde{N}_t = \int_0^t g(s)\lambda(s)\,ds$, then the mean of $\widehat{\alpha}_{n,h}(x)$ is approximated by

$$\alpha_{n,h}(x) = \alpha(x) + \frac{h^2}{2} m_{2K}\{(f\alpha)^{(2)}(x) - \alpha(x) f^{(2)}(x)\} + o(h^2).$$

The variance of $\widehat{\alpha}_{n,h}(x)$ is a $O((Th)^{-1})$

$$v_{\alpha,n,h}(x) = n^{-1}\sum_{i=1}^n \Delta_{n,i}^{-2}(nh)^{-1}\{\sigma_\alpha^2(x) + o(1)\},$$

$$\sigma_\alpha^2(x) = \kappa_2 f^{-1}(x)\Delta_n^{-1} Var(Y_t \mid X_t = x)$$
$$= \kappa_2 f^{-1}(x)\{\beta^2(x) + \gamma^2(x) g(t)\lambda(t)\}$$

and its covariances tend to zero. The process $(Th)^{1/2}(\widehat{\alpha}_{n,h} - \alpha)$ has the asymptotic variance $\kappa_2\sigma_\alpha^2(x) f_X^{-1}(x)$, at x.

The discrete part of X is $X^d(t) = \sum_{s\leq t}\gamma(X_s)\Delta N_s$ and its continuous part is $X^c(t) = \int_0^t \alpha(X_s)\,ds + \int_0^t \beta(X_s)\,dB_s - \int_0^t \gamma(X_s)\,d\widetilde{N}_s$, with variations on (t_i, t_{i+1})

$$\Delta X_i^c = \alpha(X_{t_i})\,\Delta_{n,i} + \beta(X_{t_i})\,\Delta B_{t_i} - \gamma(X_{t_i})\,\Delta_{n,i} Y(t_i)\lambda(t_i) = O_p(\Delta_{n,i}).$$

Then the sum its jumps converges to $\int_0^t E\gamma(X_s)\, g(s) d\Lambda_s$. Let $(T_i)_{1\leq i\leq N(T)}$ be the jumps of the process N. The jumps $\Delta X^d(T_i) = \gamma(X_{T_i})$ yield a consistent estimator of $\gamma(x)$, for x in $\mathcal{X}_{n,h}$

$$\widehat{\gamma}_{n,h}(x) = \frac{\sum_{1\leq i\leq N(T)} \Delta X^d(T_i) K_h(x - X_{T_i})}{\sum_{1\leq i\leq N(T)} K_h(x - X_{T_i})}$$
$$= \frac{\sum_{1\leq i\leq N(T)} \gamma(X_{T_i}) K_h(x - X_{T_i})}{\sum_{1\leq i\leq N(T)} K_h(x - X_{T_i})}.$$

The expectation of $\widehat{\gamma}_{n,h}(x)$ is approximated by the ratio of the means of the numerator and the denominator. For the numerator

$$ET^{-1}\int_0^T \gamma(X_s) K_h(x - X_s)\,dN_s = (\gamma f_N)(x) + \frac{h^2}{2} m_{2K}\{(\gamma f_N)(x)\}^{(2)} + o(h^2)$$

and, for the denominator $ET^{-1}\int_0^T K_h(x - X_s)\,dN_s = f_N(x) + \frac{h^2}{2} m_{2K} f_N^{(2)}(x) + o(h^2)$. The bias of $\widehat{\gamma}_{n,h}(x)$ is then

$$b_{\gamma,n,h}(x) = \frac{h^2}{2} m_{2K}\{f_N(x)\}^{-1}[\{\gamma(x) f_N(x)\}^{(2)} - \gamma(x) f_N^{(2)}(x)] + o(h^2),$$

also denoted $b_{\gamma,n,h}(x) = h^2 b_\gamma$. The variance of $\widehat{\gamma}_{n,h}(x)$ is deduced from the variance of the numerator

$$T^{-2}E\int_0^T K_h^2(x - X_s)\,dN_s$$
$$= (Th)^{-1}\{\kappa_2 f_N(x) + \kappa_{22}\frac{h^2}{2} f_N^{(2)}(x)\} + o(T^{-1}h_T)$$

the variance of the denominator

$$T^{-2}E\int_0^T \gamma^2(X_s)K_h^2(x-X_s)\,dN_s$$
$$= (Th)^{-1}\{\kappa_2\gamma^2(x)f_N(x)+h^2\kappa_{22}(\gamma^2 f_N)^{(2)}(x)\}+o(hT^{-1})$$

and their covariance

$$T^{-2}E\int_0^T\int_0^T \gamma(X_s)K_h(x-X_s)K_h(x-X_t)\,dN_s\,dN_t$$
$$= (Th)^{-1}\{\kappa_2\gamma(x)f_N(x)+h^2\kappa_{22}(\gamma f_N)^{(2)}(x)\}+o(hT^{-1}),$$

therefore $v_{\gamma,n,h}(x)=T^{-1}hv_\gamma(x)$ with

$$v_\gamma(x)=\kappa_{22}\{f_N(x)\}^{-1}\{(\gamma^2 f_N)(x)^{(2)}-\gamma^2(x)f_N^{(2)}(x)\}+o(hT^{-1}).$$

It follows that the process $(Th^{-1})^{1/2}(\widehat{\gamma}_n-\gamma-c_\alpha b_\gamma)$ converges weakly to a centered Gaussian process with variance function $v_\gamma(x)$ and covariances zero.

For the estimation of the variance function β of model (8.10), let

$$Z_i = Y_i-\Delta_{n,i}\widehat{\alpha}_{n,h}(X_{t_i})-\widehat{\gamma}_{n,h}(X_{t_i})\eta_i$$
$$= \Delta_{n,i}(\alpha-\widehat{\alpha}_{n,h})(X_{t_i})+\beta(X_{t_i})\varepsilon_i+(\gamma-\widehat{\gamma}_{n,h})(X_{t_i})\eta_i,$$

its conditional expectation $E(Z_i\mid X_{t_i}=x)=\Delta_{n,i}(\alpha-\alpha_{n,h})(X_{t_i})$ tends to zero and its conditional variance satisfies

$$\Delta_{n,i}^{-1}Var\{Z_i\mid X_{t_i}\}=\beta^2(X_{t_i})+o(h^4)+o((nh)^{-1})+o((Th)^{-1}).$$

An estimator of the function β is deduced for x in $\mathcal{X}_{n,h}$

$$\widehat{\beta}_{n,h}^2(x)=\frac{\sum_{1\le i\le n}\Delta_{n,i}^{-1}Z_i^2K_h(x-X_{t_i})}{\sum_{i=1}^n K_h(x-X_{t_i})}.$$

The previous approximations of the estimator $\widehat{\beta}_{n,h}$ given in Proposition 8.1 are modified, its expectation is approximated by

$$\beta_{n,h}^2(x)=n^{-1}\sum_{1\le i\le n}\Delta_{n,i}^{-1}EZ_i^2K_h(x-X_{t_i})f_N^{-1}(x)+o(h^2)$$

therefore its bias is $E\widehat{\beta}_{n,h}^2-\beta^2=b_{\beta,n,h}+o(h^2)$ with

$$b_{\beta,n,h}=\frac{h^2}{2}m_{2K}f^{-1}(x)\{(f\beta^2)^{(2)}(x)-\beta^2(x)f^{(2)}(x)\}+o(h^2).$$

Under conditions (2.1) and (3.1) for the function β in class $C_2(\mathcal{X})$, the variance of the estimator $\widehat{\beta}_{n,h}^2$ is $v_{\beta,n,h}(x)=(nh)^{-1}\{\sigma_\beta^2(x)+o(1)\}$, with

$$\sigma_\beta^2(x)=\kappa_2 f_{X_t}^{-1}(x)\Delta_n^{-2}Var(Z_t^2\mid X_t=x).$$

The normalized variance $\Delta_n^{-2} Var(Z_t^2 \mid X_t = x)$ develops as

$$E\{\Delta_n^2(\widehat{\alpha}_{n,h} - \alpha)^4(x) + \beta^4(x)\Delta_n^{-2}\varepsilon^4 + (\gamma - \widehat{\gamma}_{n,h})^4(x)\Delta_n^{-2}\eta^4 \\ + O(h^4) + O((nh)^{-1}) + O(hT^{-1})$$

where the Burkhölder-Davis-Gundy inequality implies that the order of $E\eta_i^4$ is a $O((E\eta_i^2)^2) = O(\Delta_{n,i}^2)$. Then, from the expression of the moments of the variable ε, $\sigma_\beta^2(x) = \beta^4(x)(\Delta_n^{-2}E\varepsilon^4 - 1) + o(1) = 2\beta^4(x) + o(1)$. The variance of $\widehat{\beta}_{n,h}^2$ is therefore written $v_{\beta,n,h}(x) = (nh)^{-1}v_\beta(x)$, it is a $O((nh)^{-1})$ and the process $(nh)^{1/2}(\widehat{\beta}_n - \beta - (nh^5)^{1/2}b_\beta)$ converges weakly to a centered Gaussian process with variance function v_β and covariances zero.

8.5 Continuous estimation for diffusions with jumps

In model (8.10), the estimator $\widehat{\alpha}_{T,h_T}$ of Section 8.3 is unchanged and new estimators of the functions β and γ must be defined from the continuous observation of the sample path of X. The discrete part of X is also written $X_t^d = \int_0^t \gamma(X_s)dN_s$ and the point process N is rescaled as $N_t = N_{Ts}$, with t in $[0,T]$ and s in $[0,1]$. Let

$$N_T(s) = T^{-1}N_{Ts}, \quad X_T(s) = T^{-1}X_{Ts},\ t \in [0,T],\ s \in [0,1].$$

The predictable compensator of N_T is written $\widetilde{N}_T(t) = T^{-1}\int_0^t Y_T(s)\lambda(s)\,ds$ on $[0,1]$ and it is assumed to converge uniformly on $[0,1]$ to its mean $E\widetilde{N}_T(t) = \int_0^t g(s)\lambda(s)\,ds$, in probability. Then $X_T^d(t)$ converges uniformly in probability to $\int_0^t E\gamma(X_T(s))g(s)d\Lambda(s)$. The continuous part of X is $dX_t^c = \alpha(X_t)\,dt + \beta(X_t)\,dB_t - \gamma(X_t)Y_t\lambda_t\,dt$. A consistent estimator of $\gamma(x)$, for x in $\mathcal{I}_{X,T,h}$

$$\begin{aligned}\widehat{\gamma}_{T,h}(x) &= \frac{\int_0^T K_h(x - X_s)\,dX^d(s)}{\int_0^T K_h(x - X_s)\,dN(s)}, \\ &= \frac{\int_0^T K_h(x - X_s)\gamma(X_s)\,dN(s)}{\int_0^T K_h(x - X_s)\,dN(s)},\end{aligned}$$

it is identical to the estimator previously defined for the discrete diffusion process. Its moments calculated in the continuous model (8.10) are identical to those of Section 8.4 then the process $(Th_T^{-1})^{1/2}(\widehat{\gamma}_{T,h_T} - \gamma - c_\alpha b_\gamma)$ converges weakly to a centered Gaussian process with variance function v_γ and covariances zero.

The variance function $\beta^2(X_t)$ is now estimated by smoothing the squared variations of the process

$$Z_t = X_t - X_0 - \int_0^t \widehat{\alpha}_{T,h}(X_s)\,ds \tag{8.11}$$
$$= \int_0^t (\alpha - \widehat{\alpha}_{T,h})(X_s)\,ds + \int_0^t \beta(X_s)\,dB_s + \int_0^t (\gamma - \widehat{\gamma}_{T,h})(X_s)\,dM_s.$$

For every t in $[0,T]$, its first two conditional moments are

$$E(Z_t \mid \mathcal{F}_t) = -\int_0^t b_{\alpha,T,h}(X_s)\,ds = O(h^2)$$

and

$$Var(Z_t \mid \mathcal{F}_t) = Var\{\int_0^t \widehat{\alpha}_{T,h}(X_s)ds\} + E\int_0^t \beta^2(X_s)ds$$
$$+ E\int_0^t (\gamma - \widehat{\gamma}_{T,h})^2(X_s)\,d\widetilde{N}_s$$
$$= t\int_{\mathbb{X}} \beta^2(x) f_{X_s}(x)\,dx + O((Th_T)^{-1}) + O(h_T^4).$$

Furthermore, the Burkhölder-Davis-Gundy inequality implies the existence of a constant c_4 such that

$$Var Z_t^2 = E\{\int_0^t \beta(X_s)\,dB_s\}^4 - \{t\int \beta^2(x)f(x)\,dx\}^2$$
$$\leq c_4 E\int_0^t \beta^4(X_s)\,ds = c_4 t\int \beta^2(x)f(x)\,dx.$$

The variance function $\beta^2(x)$ is then consistently estimated smoothing the process Z_t^2

$$\widehat{\beta}^2_{T,h}(x) = 2\frac{\int_0^T K_h(X_s - x)\,Z_s\,dZ_s}{\int_0^T K_h(X_s - x)\,ds}. \tag{8.12}$$

Under conditions (2.1) and (3.1) for the function β in class $C_2(\mathcal{X})$ and using the ergodicity property (2.13) for the limiting density f of the process $(X_t)_{t\in[0,T]}$, the expectation of the denominator of (8.12) is

$$T^{-1}E\int_0^T K_h(X_s - x)\,ds = T^{-1}E\int_0^T f_{X_s}(x)\,ds$$
$$+ \frac{h^2}{2}m_{2K}T^{-1}E\int_0^T f_{X_s}^{(2)}(x)\,ds + o(h^2)$$
$$= f(x) + \frac{h^2}{2}m_{2K}f^{(2)}(x) + o(h^2)$$

the expectation of the numerator is

$$\begin{aligned}2T^{-1}E\int_0^T K_h(X_s-x)Z_s\,dZ_s \\ &= 2T^{-1}\int_0^T E\int_{\mathbb{X}} K_h(u-x)\,\beta^2(u)\,f_{X_s}(u)\,du + o(h^4) \\ &= \beta^2(x)\,f(x) + \frac{h^2}{2}m_{2K}(\beta^2(x)f(x))^{(2)} + o(h^2)\end{aligned}$$

and its bias is denoted $b_{\beta,T,h} = h^2 b_\beta + o(h^2)$, with

$$b_\beta = \frac{1}{2}m_{2K}f^{-1}(x)\{(f\beta^2)^{(2)}(x) - \beta^2(x)f^{(2)}(x)\}.$$

Under conditions (2.1) and (3.1) for the function β in class C_2, the variance of the estimator $\widehat{\beta}_{T,h}$ is obtained from $E(Z_t^2 \mid X) = \int \beta^2(X_s)\,ds$, $Var(Z_t^2 \mid X) = O(t)$ and expanding

$$ET^{-2}\int_0^T\int K_h^2(x-y)Var(Z_t^2 \mid X_t = y)f_{X_t}(y)\,dy\,dt = O((hT)^{-1}),$$

it is therefore written $\sigma^2_{\beta,T,h} = (hT)^{-1}v_\beta + o((hT)^{-1})$. Then the process $(Th_T)^{1/2}(\widehat{\beta}_{T,h} - \beta - (Th_T^5)^{1/2}b_\beta)$ converges weakly to a centered Gaussian process with variance function v_β and covariances zero.

8.6 Transformations of a non-stationary Gaussian process

Consider the non-stationary processes $Z = X \circ \Phi$, where X is a stationary Gaussian process with covariance $R(x,y) = E(X_x X_y)$ and Φ is a monotone function $C_1([0,1])$ with $\Phi(0) = 0$ and $\Phi(1) = 1$. The transform is expressed as $\Phi(x) = v^{-1}(1)v(x)$ with respect to the integrated singularity function of the covariance $r(x,x)$, $v(x) = \int_0^x \xi(u)\,du$. Conversely, $\int_0^x \xi(u)\,du = c_\xi\Phi(x)$ with $c_\xi = \int_0^1 \xi(u)\,du$. A direct estimator of the regularity function ξ is obtained by smoothing the estimator $\widehat{\Phi}_n(x)$ defined by (1.12)

$$\begin{aligned}\widehat{\xi}_{n,h}(x) &= V_n(1)\int_0^1 K_h(x-y)\,d\widehat{\Phi}_n(y) \\ &= \int_0^1 K_h(x-y)\,d\widehat{v}_n(y).\end{aligned}$$

The expectation of $\widehat{\xi}_{n,h}(x)$ is $\xi_{n,h}(x) = \int_0^1 K_h(x-y)\,dv(y)$ and the process $(\widehat{\xi}_{n,h} - \xi_{n,h})(x) = \int_0^1 K_{h_n}(x-y)d(\widehat{\Phi}_n - \Phi)(y)$ is uniformly consistent, since an integration by parts implies

$$\begin{aligned}\|\int_0^1 K_{h_n}(s-y)\, d(\widehat{\Phi}_n - \Phi)(y)\| &\leq \|\widehat{\Phi}_n - \Phi\| \int_0^1 |dK_{h_n}(s-y)| \\ &\quad + \sup |K_{h_n}|\, \|\widehat{\Phi}_n - \Phi\| \\ &\leq (\sup |K| + \int |dK(z)|)\, h_n^{-1}\, \|\widehat{\Phi}_n - \Phi\|\end{aligned}$$

which converges to zero in probability, by the weak convergence of $n^{1/2}\|\widehat{\Phi}_n - \Phi\|$. The process $n^{1/2}(\widehat{v}_n - v)$ converges weakly to the process $\sqrt{2}\int_0^x v(y)dW(y)$ where W is a Gaussian process with mean zero and covariances $x \wedge y$ at (x,y), then the covariance of the limit of $n^{1/2}(\widehat{v}_n - v)$ is $2\int_0^{x\wedge y} v^2(y)\, dy$ at $x \neq y$. The limiting variance of $\widehat{\xi}_{n,h}(x)$ is

$$\begin{aligned}E\{\int_0^1 K_h(x-y)\, d(\widehat{v}_n - v)(y)\}^2 &= E\int_0^1 K_h^2(x-y)\, dVar(\widehat{v}_n - v)(y) \\ &\quad + E\int_0^1\int_0^1 K_h(x-y)K_h(x-u)\, dCov\{(\widehat{v}_n - v)(y), (\widehat{v}_n - v)(u) \\ = O(n^{-1}\int_0^1 K_h^2(x-y)v^2(y)\, dy) &= O((nh)^{-1})\end{aligned}$$

The convergence rate of the process $\widehat{\xi}_{n,h}$ is therefore $(nh_n)^{1/2}$ and the finite dimensional distributions of $(nh_n)^{1/2}(\widehat{\xi}_{n,h} - \xi_{n,h})$ converge to those of a Gaussian process with mean zero, as normalized sums of the independent variables defined as the weighted quadratic variations of the increments of Z. The covariances of $(nh_n)^{1/2}(\widehat{\xi}_{n,h} - \xi_{n,h})$ are zero except on the interval $[-h_n, h_n]$ where they are bounded, hence the covariance function converges to zero. The quadratic variations of $\widehat{\xi}_{n,h}$ satisfy a Lipschitz property of moments

$$\begin{aligned}&E|(\widehat{\xi}_{n,h} - \xi_{n,h})(x) - (\widehat{\xi}_{n,h} - \xi_{n,h})(y)|^2 \\ &= 2n^{-1}|\int_0^1 \{K_h^2(x-u) - K_h^2(y-u)\}v^2(u)\, du|\end{aligned}$$

it is then a $O((nh_n^3)^{-1}|x-y|^2)$ for $|x-y| \leq 2h_n$. It follows that the process $(nh_n)^{1/2}(\widehat{\xi}_{n,h} - \xi_{n,h})$ converges weakly to a continuous process with mean zero and variance function $2v^2$ and covariances zero.

The singularity function of the spatial covariance of a Gaussian process Z is estimated by smoothing the estimator of the integrated spatial transform of Z on $[0,1]^3$, the convergence rate of the estimator is then $(nh^3)^{1/2}$.

8.7 Exercises

(1) Calculate the moments of the estimators for the continuous process (8.6) and write the necessary ergodic conditions for the convergences in this model.
(2) Calculate the bias and variance of derivatives of the estimators of functions α, β and γ in the stochastic differential equations model (8.10).
(3) Prove Proposition 8.2.

Chapter 9

Applications to time series

Let $(\mathbb{X}, \|\cdot\|)$ be a metric space and $(X_t)_{t\in\mathbb{N}}$ be a time series defined on $\mathbb{X}^{\mathbb{N}}$ by its initial value X_0 and a recursive equation $X_t = m(X_{t-p}, \dots, X_{t-1}) + \varepsilon_t$ where m is a parameric or nonparametric function defined on $\mathbb{X}^p$ for some $p > 1$ and $(\varepsilon_t)_t$ is a sequence of independent noise variables with mean zero and variance σ^2, such that for every t, ε_t is independent of $(X_{t-p}, \dots, X_{t-1})$.

The stationarity of a time series is a property of the joint distribution of consecutive observations. The weak stationarity is defined by a constant mean μ and a stationary covariance function

$$\rho_{s,t} = Cov(X_s, X_t) = Cov(X_0, X_{t-s}), \text{ for every } s < t.$$

The series $(X_t)_t$ is strong stationary if the distributions of the sequences $(X_{t_1}, \dots, X_{t_k})$ and $(X_{t_1-s}, \dots, X_{t_k-s})$ are identical for every sequence $(t_1, \dots, t_k, s)$ in $\mathbb{N}^{k+1}$. The nonparametric estimation of the mean and the covariances is therefore useful for modelling the time series. The moving average processes are stationary, they are defined as linear combinations of past and present noise terms such as the MA(q) process $X_t = \varepsilon_t + \sum_{k=1}^q \theta_k \varepsilon_{t-k}$, with independent variables ε_j such that $E\varepsilon_j = 0$ and $Var\varepsilon_j = \sigma^2$, for every integer j. The variance of X_t is $\sigma_q^2 = \sigma^2\big(\sum_{k=1}^q \theta_k^2 + 1\big)$ and it is supposed to be finite. The covariance of X_s and X_t such that $0 < t - s < q$ is $Cov(X_s, X_t) = \sigma^2\big(\sum_{k=t-s}^q \theta_k + \sum_{k=t-s+1}^{(t-s+q)\wedge q} \theta_k^2\big)$, it only depends on the difference $t - s$. The moving average processes with $|\theta| < 1$ are reversible and the process X_t can be expressed as an auto-regressive process, sum of ε_t and an infinite combination of its past values. Generally, an AR process is not stationary.

In nonstationary series, a nonstationarity may be due to a smooth trend or regular and deterministic seasonal variations, to discontinuities or to a continuous change-points. A transformation such as differencing a stochastic linear trend reduces the nonstationarity of the series, other classical transformations are the square root or power transformations for data with increasing variance. Periodic functions of the mean can be estimated after the identification of the period and nonparametric estimator is proposed in Section 9.2. Change-points of nonparametric regressions in time or at thresholds of the series are stronger causes of non regularity and several phases of the series must be considered separately, with estimation of their change-points. Their estimators are studied in Section 9.5.

9.1 Nonparametric estimation of the mean

The simplest nonparametric estimators for the mean of a stationary process are the moving average estimators

$$\widehat{\mu}_{t,k} = \frac{1}{k+1} \sum_{i=0}^{k} X_{t-i},$$

for a lag k up to t. The transformed series is $X_t - \widehat{\mu}_{t,k} = \frac{k}{k+1} X_t - \frac{1}{k+1} \sum_{i=1}^{k} X_{t-i}$ and it equals $(X_t - X_{t-1})/2$ for $k = 2$. A polynomial trend is estimated by minimizing the empirical mean squared error of the model, then the transformed series $X_t - \widehat{\mu}_{t,k}$ is expressed by the means of moving average of higher order, according to the degree of the polynomial model.

Consider the auto-regressive process with nonparametric mean

$$X_t = \mu_t + \alpha X_{t-1} + \varepsilon_t, \; t \in \mathbb{N}, \tag{9.1}$$

with an independent sequence of independent errors $(\varepsilon_t)_t$ with mean zero and variance σ^2. With $\alpha \neq 1$, its mean μ_t may be written $(1-\alpha)m_t$, with an unknown function m_t and the solution X_t of Equation (9.1) is

$$X_t = \sum_{k=0}^{t-1} \mu_{t-k} \alpha^k + \alpha^t X_0 + \sum_{k=1}^{t} \alpha^k \varepsilon_{t-k}.$$

With a mean and an initial value zero, the covariance of X_s and X_t is $\rho_{s,t} = \sigma^2 \sum_{k=1}^{s \wedge t} \alpha^{2k}$ and it is not stationary. The asymptotic behaviour of the process X changes as the mean crosses the threshold value 1. For $\alpha = 1$, the model is the classical nonparametric regression model.

The parameters of the auto-regressive series AR(1) $X_t = \mu + \alpha X_{t-1} + \varepsilon_t$, with $\alpha \neq 1$, are estimated by

$$\begin{aligned}
\widehat{\mu}_t &= (1-\widehat{\alpha}_t)\bar{X}_t + t^{-1}(\widehat{\alpha}_t X_t - X_0),\\
\widehat{m}_t &= (1-\widehat{\alpha}_t)^{-1}\widehat{\mu}_t,\\
\widehat{\alpha}_t &= \frac{\sum_{k=1}^t (X_{k-1}-\widehat{m}_t)(X_k-\widehat{m}_t)}{\sum_{k=1}^t (X_{k-1}-\widehat{m}_t)^2}, \qquad (9.2)\\
\widehat{\sigma}_t^2 &= \frac{1}{t}\sum_{k=1}^t \{X_k - (1-\widehat{\alpha}_t)\bar{X}_t - \widehat{\alpha}_t X_{k-1}\}^2.
\end{aligned}$$

For $|\alpha| \neq 1$, $\widehat{\mu}_t = (1-\widehat{\alpha}_t)\bar{X}_t + O_p(t^{-1})$ and $\widehat{m}_t = \bar{X}_t$. For $\alpha = 1$, the parametrization $\mu = (1-\alpha)m$ is meaningless and the mean is estimated by $\widehat{\mu}_t = t^{-1}\sum_{k=1}^t (X_k - X_{k-1})$. The estimators are consistent and asymptotially Gaussian, with different normalization sequences for the three domains of α ($\alpha < 1$, $\alpha = 1$, $\alpha > 1$). In the AR(p) model $X_t = \mu_t + \sum_{j=1}^p \alpha_j X_{t-j} + \varepsilon_t$, similar estimators are defined for the regression parameters α_j

$$\widehat{\alpha}_{j,t} = \frac{\sum_{k=j}^t (X_{k-j}-\widehat{m}_t)(X_k-\widehat{m}_t)}{\sum_{k=j}^t (X_{k-j}-\widehat{m}_t)^2}$$

and the variance is estimated by the mean squared estimation error.

In model (9.1) with a nonparametric mean function $\mu_t = (1-\alpha)m_t$, $X_t - m_t = \alpha(X_{t-1} - m_t) + \varepsilon_t$, then the estimator (9.2) of α is modified by replacing $\widehat{m}_k = \bar{X}_k$ by a local moving average mean or by a local mean

$$\widehat{m}_k = \frac{\sum_{j=0}^t K_h(j-k)X_j}{\sum_{j=1}^t K_h(j-k)}$$

for every k, and the estimator of α becomes

$$\widehat{\alpha}_t = \frac{\sum_{k=1}^t (X_{k-1}-\widehat{m}_k)(X_k-\widehat{m}_k)}{\sum_{k=1}^t (X_{k-1}-\widehat{m}_k)^2}.$$

Finally, the function μ_t is estimated by $(1-\widehat{\alpha}_t)\widehat{m}_t$ or by smoothing $X_t - \widehat{\alpha}_t X_{t-1}$

$$\widehat{\mu}_{t,h,k} = t^{-1}\sum_{j=0}^t K_h(j-k)(X_j - \widehat{\alpha}_j X_{j-1})$$

and the estimator of σ^2 is still defined by (9.2). The asymptotic distributions are modified as a consequence of the asymptotic behaviour of $\widehat{m}_k$, with mean tending to m_k and variance converging to a finite limit. As h tends to zero, the weak convergence to centered Gaussian variables of

$t^{1/2}(\widehat{m}_t - m_t)$, when $|\alpha| < 1$, and $t\alpha^{-t}(\widehat{m}_t - m_t)$, when $|\alpha| > 1$, follows from martingale properties of the time series which imply its ergodicity and a mixing property (Appendix D). If $|\alpha| \neq 1$

$$\widehat{\alpha}_t - \alpha = \frac{\sum_{k=1}^t (X_{k-1} - \widehat{m}_{k-1})((1-\alpha)(m_k - \widehat{m}_k) + \varepsilon_k)}{\sum_{k=1}^t (X_{k-1} - \widehat{m}_k)^2}$$

it is therefore approximated in the same way as in model AR(1) and it converges weakly with the same rate as in this model.

When Equation (9.1) is defined by a regular parametrization of the mean $\mu_t = (1-\alpha)m_\theta(t)$ for $|\alpha| \neq 1$, the minimization of squared estimation error $\|\widehat{\varepsilon}^2_{(t)}\|^2_t = \sum_{k=1}^t \widehat{\varepsilon}^2_k = \sum_{k=1}^t \{X_k - (1-\widehat{\alpha}_t)\bar{X}_t - \widehat{\alpha}_t X_{k-1}\}^2$ yields estimators of the parameters α and θ for identically distributed error variables ε_k. If the variance of ε_k is $\sigma^2_k(\theta)$, maximum likelihood estimators minimize $\sum_k \sigma_k^{-1}\varepsilon_k^2$. The robustness and the bias of the estimators in false models have been studied for generalized exponential distributions, the same methods are used in models for time series.

In a nonparametric regression model

$$X_t = m(X_{t-1}) + \varepsilon_t \tag{9.3}$$

with an initial random value X_0 and with independent and identically distributed errors ε_t with mean zero and variance σ^2, let F be the continuous distribution function of the variables ε_t, and f its density. The nonparametric estimator of the function m is still

$$\widehat{m}_{t,h}(x) = \frac{\sum_{k=1}^t K_h(x - X_{k-1})X_k}{\sum_{k=1}^t K_h(x - X_{k-1})}.$$

It is uniformly consistent under the ergodicity condition

$$\frac{1}{t}\sum_{k=1}^t \varphi(X_k, X_{k-1}) \to \int\int \varphi(x,y)F(dx - m(y))\,d\pi(y)$$

with the invariant measure π of the process and for every continuous and bounded function φ on $\mathbb{R}^2$. Conditions on the function m and the independence of the error variables ε_i ensure the ergodicity, then the process $(th)^{1/2}(\widehat{m}_{t,h} - m)$ converges weakly to a continuous centered Gaussian process with covariances zero and variance $\kappa_2 f^{-1}(x)Var\{X_k \mid X_{k-1}\}$, where $Var\{X_k \mid X_{k-1}\} = \sigma^2$. In model (9.3) with a functional variance, the results of Section 3.6 apply.

The observation of series in several groups or in distinct time intervals may introduce a group or time effect similar to population effect in regression samples and sub-regression functions may necessary as in Section 5.6.

9.2 Periodic models for time series

Let $(X_t)_{t\in\mathbb{N}}$ be a periodic auto-regressive time series defined by X_0 and

$$X_t = \psi(t) + \sum_{i=1}^{p} \alpha_p X_{t-p} + \varepsilon_t \tag{9.4}$$

where $|\alpha| < 1$ and ψ is a periodic function defined in $\mathbb{N}$ with period τ, $\psi(t) = \psi(t + k\tau)$, for every integers t and k. Let $\alpha = (\alpha_1 \ldots, \alpha_p)$ and $X_{(p),t} = (X_{t-1}, \ldots, X_{t-p})$. As $\psi(t) = E(X_t - \alpha^T X_{(p),t})$, the value of the function ψ at t is estimated by an empirical mean over the periods, with a fixed parameter value α. Assuming that K periods are observed and $T = K\tau$ values of the series are observed, the function ψ is estimated as a mean over the K periods of the remainder term of the auto-regressive process. For every t in $\{1, \ldots, \tau\}$

$$\widehat{\psi}_{K,\alpha}(t) = \frac{1}{K} \sum_{k=0}^{K-1} (X_{t+k\tau} - \alpha^T X_{(p),t+k\tau}) \tag{9.5}$$

and the parameter vector is estimated by minimizing the mean squared error of the model

$$l_K(\alpha) = \frac{1}{T} \sum_{t=1}^{T} \{X_t - \widehat{\psi}_{K,\alpha}(t) - \alpha^T X_{(p),t}\}^2.$$

The components of the first two derivatives of l_K are

$$\begin{aligned}
\dot{l}_{T,K,t} &= -\frac{2}{T} \sum_{t=1}^{T} \{X_t - \widehat{\psi}_{K,\alpha}(t) - \alpha^T X_{(p),t}\}\{\frac{\partial \widehat{\psi}_{K,\alpha}}{\partial \alpha}(t) + X_{(p),t}\} \\
&= \frac{2}{T} \sum_{t=1}^{T} \{X_t - \widehat{\psi}_{K,\alpha}(t) - \alpha^T X_{(p),t}\}\{\frac{1}{K} \sum_{k=0}^{K-1} X_{(p),t+k\tau} - X_{(p),t}\}, \\
\ddot{l}_{T,K,t} &= \frac{2}{T} \sum_{t=1}^{T} \{\frac{1}{K} \sum_{k=0}^{K-1} (X_{(p),t+k\tau} - X_{(p),t})\}^{\otimes 2}.
\end{aligned}$$

The vector α is estimated by $\widehat{\alpha}_T = \arg\min_{\alpha \in]-1,1[^d} l_{T,K,t}(\alpha)$. For the first order derivative, $T^{1/2}\dot{l}_{T,K,t}(\alpha_0)$ converges weakly to a centered limiting distribution and the second order derivative $\ddot{l}_{T,K,t}$ converges in probability to a positive definite matrix $E\ddot{l}_{T,K,t}$ which does not depend on α. Then the estimator of α satisfies $T^{1/2}(\widehat{\alpha}_{T,K,t} - \alpha_0) = \ddot{l}_{T,K,t}^{-1} T^{1/2}\dot{l}_{T,K,t}(\alpha_0) + o(1)$. The estimator $\widehat{\alpha}_T$ is consistent and its weak convergence rate is $T^{1/2}$, if all components of the vector α have a norm smaller than 1. The function ψ is then consistently estimated by $\widehat{\psi}_K = \widehat{\psi}_{K,\widehat{\alpha}_T}$ and, for every t in $\{1,\ldots,\tau\}$, the weak convergence rate of the estimator $\widehat{\psi}_K(t)$ is $K^{1/2}$.

The true period of the function ψ was supposed to be known. With an unknown period, the estimators $\widehat{\psi}_K$ and $\widehat{\alpha}_T$ depend on the parameter τ and it is consistently estimated by $\widehat{\tau}_T = \arg\min_{\tau\leq T} l_{[T/\tau]}(\widehat{\alpha}_{T,\tau})$.

If the function ψ is parametric, its parameters vector θ is estimated by minimizing the mean squared error between $\widehat{\psi}_K$ and ψ_θ, $\frac{1}{T}\sum_{t=1}^{T}\{\widehat{\psi}_K(t) - \psi_\theta(t)\}^2$. As a minimum distance estimator, the estimator $\widehat{\theta}_K$ is consistent and $T^{1/2}(\widehat{\theta}_T - \theta)$ converges weakly to a centered Gaussian variable.

The trigonometric series with independent noise are a combination of periodic sinus and cosinus functions

$$\begin{aligned}
X_t &= \sum_{j=1}^{r} M\{\cos(w_j t + \Phi_t) + \sin(w_j t + \Phi_t)\} + \varepsilon_t \\
&= \sum_{j=1}^{r}\{A_j \cos(2\pi w_j t) - B_j \sin(2\pi w_j t)\} + \varepsilon_t
\end{aligned}$$

where $(w_j)_{j=1,\ldots,r}$ are frequencies $w_j = jt^{-1}$, $A_j = M\cos\Phi_t$, $B_j = M\sin\Phi_t$ such that $A_j^2 + B_j^2 = M^2$ is the magnitude of the series, for $j = 1,\ldots,r$, and Φ_t its phases. The estimators of the parameters are defined from the Fourier series, for $j = 1,\ldots,r$

$$\begin{aligned}
\widehat{A}_{tj} &= 2n^{-1}\sum_{k=1}^{t} X_t \cos(2\pi kj/t), \\
\widehat{B}_{tj} &= 2n^{-1}\sum_{k=1}^{t} X_t \sin(2\pi kj/t), \\
\widehat{M}_t &= r^{-1}\sum_{j=1}^{r}(\widehat{A}_{tj}^2 + \widehat{B}_{tj}^2)^{1/2}.
\end{aligned}$$

9.3 Nonparametric estimation of the covariance function

The classical estimator for estimating the covariances function in a stationary model is similar to the moving average for the mean, with a lag $k \geq 1$ between variables X_i and X_{i-k}, for every $i \geq 1$,

$$\widehat{\rho}_{k,t} = (t-k)^{-1}\sum_{i=k+1}^{t}(X_i - \bar{X}_t)(X_{i-k} - \bar{X}_t).$$

In the auto-regressive model AR(1) with independent errors with mean zero and variance σ^2, for $k \geq 1$, the variable X_k is expressed from the initial value as

$$X_k - m = \alpha^k(X_0 - m) + S_{k,\alpha}, \text{ where } S_{k,\alpha} = \sum_{j=1}^{k} \alpha^{k-j}\varepsilon_j = \sum_{j=0}^{k-1} \alpha^j \varepsilon_{k-j}.$$

Let B be the standard Brownian motion, if $|\alpha| < 1$ the process $S_{[ns],\alpha}$ defined up to the integer part of ns converges weakly to $\sigma B\{(1-\alpha^2)^{1/2}\}^{-1}$. If $\alpha = 1$, the process $n^{-1/2}S_{[ns],1}$ converges weakly to σB, and if $|\alpha| > 1$ the process $\alpha^{-[ns]}S_{[ns],\alpha}$ converges weakly to $\sigma B\{(\alpha^2-1)^{1/2}\}^{-1}$. The independence of the error variables ε_j implies

$$E(X_k - m)(X_{k+s} - m) = \alpha^{2k+s} Var X_0 + Cov(S_{k,\alpha}, S_{k+s,\alpha}), \quad (9.6)$$

$$Cov(S_{k,\alpha}, S_{k+s,\alpha}) = E(\sum_{j=1}^{k} \alpha^{k-j}\varepsilon_j)^2 = \sigma^2 \sum_{j=1}^{k} \alpha^{2(k-j)},$$

so $E(X_k - m)(X_{k+s} - m) = \alpha^{2k+s} Var X_0 + Var S_{k,\alpha}$ and the covariance function of the series is not stationary. The estimator (9.2) of the variance σ^2 is defined as the empirical variance of the estimator of the noise variables which are identically distributed and independent. In the same way, the covariance is estimated by

$$\widehat{\rho}_{t,k} = \frac{1}{t-k} \sum_{i=k+1}^{t} \{X_i - \widehat{m}_t - \widehat{\alpha}_t(X_{i-1} - \widehat{\alpha}_t)\}\{X_{i-k} - \widehat{m}_t - \widehat{\alpha}_t(X_{i-k-1} - \widehat{\alpha}_t)\},$$

the estimators $\widehat{\sigma}_t^2$ and $\widehat{\rho}_{t,k}$ are consistent (Pons 2008). The estimators are defined in the same way in an auto-regressive model of order p, with a scalar products $\widehat{\alpha}_t^T X_{i-1}$ and $\widehat{\alpha}_t^T X_{i-k-1}$ for p-dimensional variables X_{i-1} and X_{i-k-1}. In model (9.1), the expansion (9.6) of the variables centered by the mean function is not modified and the covariance $E(X_k - m_k)(X_{k+s} - m_{k+s})$ has the same expression depending only on the variances of the initial value and $S_{k,\alpha}$, and on α and the rank of the observations. In auto-regressive series with deterministic models of the mean, the covariance estimator is modified by the corresponding estimator of the mean. In model (9.3), the covariance estimator becomes

$$\widehat{\rho}_{t,k} = \frac{1}{t-k} \sum_{i=k+1}^{t} \{X_i - \widehat{m}_{t,h}(X_i)\}\{X_{i-k} - \widehat{m}_{t,h}(X_{i-k})\}$$

and the estimators are consistent.

9.4 Nonparametric transformations for stationarity

In the nonparametric regression model (9.3), $X_t = m(X_{t-1}) + \varepsilon_t$ with an initial random value X_0 and with independent and identically distributed errors ε_t with mean zero and variance σ^2, the covariance between X_k and X_{k+l} is $\rho_{t,k,l} = E\{X_k m^{*l}(X_k)\} - EX_k Em^{*l}(X_k)$, with $E\{X_k m(X_{k+l-1})\} = E\{X_k m^{*l}(X_k)\}$, where m^{*l} is the composition of l functions m. The nonstationarity of $\rho_{t,k,l}$ does not allow to estimate it using empirical means and it is necessary to remove the functional mean μ_t before studying the covariance of the series. The centered series

$$Y_t = X_t - \widehat{m}_t(X_{t-1}) = m(X_{t-1}) - \widehat{m}_t(X_{t-1}) + \varepsilon_t$$

has a conditional expectation equal to minus the bias of the estimator $\widehat{m}_t$

$$E(Y_t \mid X_{t-1}) = -\frac{h^2}{2}\{(mf_{X_{t-1}})^{(2)} - mfX_{t-1}^{(2)}\}(X_{t-1})m_{2K}f_{X_{t-1}}^{-1}(X_{t-1})$$

and it is negligeable as t tends to infinity and h to zero. The time series Y_t is then asymptotically equivalent to a random walk with a variance parameter σ^2. The main transformations for nonstationary series (9.3) with a constant variance is therefore its centering.

With a varying variance function

$$E\varepsilon_i^2 = \sigma_i^2 = Var(X_i \mid X_{i-1}),$$

the estimator of the mean function of the series has to be weighted by the inverse of the square root of the nonparametric estimator of the variance at X_i, where

$$\widehat{\sigma}^2_{t,h,\delta}(x) = \frac{\sum_{i=1}^t \{Y_i - \widehat{m}_{t,h}(X_i)\}^2 K_\delta(x - X_i)}{\sum_{i=1}^n K_\delta(x - X_i)},$$

as in Section 3.6, the estimator of the regression function is

$$\widehat{m}_{w,t,h}(x) = \frac{\sum_{i=1}^t \widehat{w}_{t,h,\delta}(X_i) Y_i K_h(x - X_i)}{\sum_{i=1}^n \widehat{w}_{t,h,\delta}(X_i) K_h(x - X_i)}$$

and the stationary series for (9.3) is $Y_i = X_i - \widehat{m}_{w,t,h}(X_{i-1})$. A model for non independent stationary terms ε_t can then be detailed.

9.5 Change-points in time series

A change-point in a time series may occur at an unknown time τ or at an unknown threshold η of the series. In both cases, X_t splits into two processes at the unknown threshold

$$X_{1,t} = X_t I_t \text{ and } X_{2,t} = X_t(1 - I_t)$$

with $I_t = 1\{X_t \leq \eta\}$ for a model with a change-point at a threshold of the series and $I_t = 1\{t \leq \tau\}$ in a model with a time threshold. The p-dimensional parameter vector α is replaced by two vectors α and β. Both change-points models are written equivalently, with a time change-point or a series change-points

$$\tau_\eta = \sup\{t; X_t \leq \eta\},\ \eta_\tau = \sup\{x; (X_s)_{s\in[0,\tau]} \leq x\}. \tag{9.7}$$

With a change-point, the auto-regressive model AR(p) is modified as

$$X_t = \mu_1 I_t + \mu_2(1 - I_t) + \alpha^T X_{1,t} + \beta^T X_{2,t} + \varepsilon_t \tag{9.8}$$

where $X_t = \mu + \alpha^T X_{1,t} + \beta^T X_{2,t} + \varepsilon_t$ with $X_{1,t} = X_t I_t$ and $X_{2,t} = X_t(1-I_t)$ for a model without change-point in the mean. Considering first that the change-point is known, the parameters are μ, or μ_1 and μ_2, α, β and σ^2. As t tends to infinity, a change-point at an integer time τ is denoted $[\gamma t]$ and sums of variables up to τ are increasing with t. For the auto-regressive process of order 1 with a change-point in time, this equation yields a two-phase sample-path

$$X_{t,\alpha} = m_\alpha + \alpha^t(X_0 - m_\alpha) + \sum_{k=1}^{t} \alpha^{t-k}\varepsilon_k,\ t \leq \tau,$$

$$X_{t,\beta} = m_\beta + \beta^{t-\tau}(X_{\tau,\alpha} - m_\alpha) + \sum_{k=1}^{t-\tau} \beta^{t-\tau-k}\varepsilon_{k+\tau},\ t > \tau,$$

or $m_\beta = \mu(1-\beta)^{-1}$. With $\alpha = 1$, $X_{t,\alpha} = X_0 + (t-1)\mu + \sum_{k=1}^{t} \varepsilon_k$ and with $\beta = 1$ and $t > \tau$, $X_{t,\beta} = X_{\tau,\alpha} + (t-k-1)\mu + \sum_{k=1}^{t-\tau} \varepsilon_{k+\tau}$. Let θ be the vector of parameters $\alpha, \beta, m_\alpha, m_\beta, \gamma$. The time τ corresponds either to a change-point of the series or a stopping time defined by (9.7) for a change-point at a threshold of the process X and the indicator I_k relative to an unknown threshold τ_η of X_{t-k} is denoted $I_{k,\tau}$.

$$\widehat{\alpha}_{t,\tau} = \frac{\sum_{k=1}^{t}(I_{k-1,\tau}X_{k-1} - \widehat{m}_{\alpha,\tau})(I_{k,\tau}X_k - \widehat{m}_{\alpha,\tau})}{\sum_{k=1}^{\tau}(I_{k-1}X_{k-1} - \widehat{m}_{\alpha,\tau})^2},$$

$$\widehat{\beta}_{t,\tau} = \frac{\sum_{k=1}^{t}((1 - I_{k-1,\tau})X_{k-1} - \widehat{m}_{\beta,t})((1 - I_{k,\tau})X_k - \widehat{m}_{\beta,t})}{\sum_{k=1}^{t}((1 - I_{k-1,\tau})X_{k-1} - \widehat{m}_{\beta,t})^2},$$

where the estimators of $m_\alpha = (1-\alpha)^{-1}\mu$ and $m_\beta = (1-\beta)^{-1}\mu$ are equivalent to

$$\widehat{m}_{\alpha,\tau} = \bar{X}_\tau,$$

$$\widehat{m}_{\beta,\tau+k} = \bar{X}_{\tau,k} = k^{-1}\sum_{j=1}^{k} X_{\tau+j}, \quad \text{for } t = \tau + k \geq \tau,$$

and $\widehat{\mu}_t = \bar{X}_\tau - \widehat{\alpha}_{t,\tau}\bar{X}_\tau - \widehat{\beta}_{t,\tau}\bar{X}_{\tau,t}$ if $|\alpha|$ and $|\beta| \neq 1$.

The estimator of the change-point parameter minimizes with respect to τ the mean squared error of estimation. For $t > \tau$, consider the estimation errors $\widehat{\varepsilon}_{\tau,k} = X_k - \widehat{m}_{\widehat{\alpha}_{t,\tau},\tau} - \widehat{\alpha}_{t,\tau} X_{k-1}$ if $k \leq \tau$ and $\widehat{\varepsilon}_{t,k} = X_k - \widehat{m}_{\widehat{\beta}_{t,\tau},t} - \widehat{\beta}_{t,\tau} X_{k-1}$ if $k > \tau$. The variance σ^2 and the change-point parameter are estimated by

$$\sigma_t^2(\theta) = \tau^{-1} \sum_{k=1}^{\tau} \widehat{\varepsilon}_{\tau,k}^2 + (t-\tau)^{-1} \sum_{k=\tau+1}^{t} \widehat{\varepsilon}_{t,k}^2,$$

$$\widehat{\gamma}_t = \arg \min_{\tau \in [0,t]} \widehat{\sigma}_t^2(\tau).$$

The change-point estimator is approximated by

$$\widehat{\gamma}_t = \arg \min_{\tau \in [0,t]} t^{1/2} \{ \frac{1}{\tau} \sum_{k=\tau_0+1}^{\tau} (X_k - \mu_\alpha - \alpha X_{k-1})^2$$

$$- \frac{1}{t-\tau} \sum_{k=\tau_0+1}^{\tau} (X_k - \mu_\beta - \beta X_{k-1})^2 - \gamma_0 \} + o_p(1),$$

$\widehat{\gamma}_t - \gamma_0$ is independent of the estimators of the parameter vector $\widehat{\xi}_t$ of the regression and all estimators converge weakly to limits bounded in probability.

Consider the model (9.8) of order 1 with a change-point at a threshold η of the series, with the equivalence (9.7) between the chronological change-point model and the model for a series crossing the threshold η at consecutive random stopping times $\tau_1 = \inf\{k \geq 0 : I_k = 0\}$ and $\tau_j = \inf\{k > \tau_{j-1} : I_k = 0\}$, $j \geq 1$. The series have similar asymptotic behaviour starting from the first value of the series which goes across the threshold η at time $s_j = \inf\{k > \tau_{j-1} : I_k = 1\}$ after τ_{j-1}. The estimators of the parameters in the first phase of the model are restricted to the set of random intervals $[s_j, \tau_j]$ where X_t stands below η, for the second phase the observations are restricted to the set of random intervals $]\tau_{j-1}, s_j[$ where X remains above η. The time τ_j are stopping times of the series defined for $t > s_{j-1}$ by

$$X_t = \begin{cases} m_\alpha + S_{s_{j-1},t-s_{j-1},\alpha} + o_p(1), & \text{if } |\alpha| < 1, \\ X_{s_{j-1}} + (t - s_{j-1} - 1)\mu + S_{s_{j-1},t-s_{j-1},1}, & \text{si } \alpha = 1, \\ m_\alpha + \alpha^{t-s_{j-1}}(X_{s_{j-1}-1} - m_\beta) + S_{s_{j-1},t-s_{j-1},\alpha}, & \text{if } |\alpha| > 1 \end{cases}$$

and the s_j are stopping times defined for $t > \tau_{j-1}$ by

$$X_t = \begin{cases} m_\beta + S_{\tau_{j-1},t-\tau_{j-1},\beta} + o_p(1), & \text{if } |\beta| < 1, \\ X_{\tau_{j-1}} + (t - \tau_{j-1} - 1)\mu + S_{\tau_{j-1},t-\tau_{j-1},1}, & \text{if } \beta = 1, \\ m_\beta + \beta^{t-\tau_{j-1}}(X_{\tau_{j-1}} - m_\alpha) + S_{\tau_{j-1},t-\tau_{j-1},\beta} & \text{if } |\beta| > 1, \end{cases}$$

The sequences $t^{-1}\tau_j$ and $t^{-1}s_j$ converge to the corresponding stopping times of the limit of X_t as t tends to infinity. The partial sums are therefore defined as sums over indices belonging to countable union of intervals $[s_j, \tau_j]$ and $]\tau_j, s_{j+1}[$, respectively, for the two phases of the model. Theirs limits are deduced from integrals on the corresponding sub-intervals, instead of sums of the errors on the interval (τ, τ_0). The estimators of the parameters are still expressions of their partial sums. The results generalize to processes of order p with a possible change-point in each p component. The estimators and their weak convergences are detailed in Pons (2009). Change-points in nonparametric models for time series are estimated by replacing the estimators of the parameters by those of the functions of the models and only the expression of the errors ε_k determines its estimator.

With a change-point at an unknown time τ_0 in the nonparametric model (9.3), it is written $X_t = I_{\tau,t}m_1(X_{t-1}) + (1 - I_{\tau,t})m_2(X_{t-1}) + \sigma\varepsilon_t$. For every x of I_X, the two regression functions are estimated using a kernel estimator with the same bandwidth h for m_1 and m_2

$$\widehat{m}_{1,t,h}(x,\tau) = \frac{\sum_{i=1}^t K_h(x - X_i)(1 - I_{\tau,i})Y_i}{\sum_{i=1}^t K_h(x - X_i)(1 - I_{\tau,i})},$$

$$\widehat{m}_{2,t,h}(x,\tau) = \frac{\sum_{i=1}^t K_h(x - X_i)I_{i,\tau}Y_i}{\sum_{i=1}^t K_h(x - X_i)I_{i,\tau}}.$$

The behaviour of the estimators $\widehat{m}_{1,t,h}$ and $\widehat{m}_{2,t,h}$ is the same as in the model where τ_0 is known, and it is the behaviour described in Section 9.1. The variance σ^2 is estimated by

$$\widehat{\sigma}^2_{\tau,t,h} = t^{-1}\sum_{i=1}^t \{Y_i - (I_{\tau,i})\widehat{m}_{1,t,h}(X_i,\tau) - (1 - I_{\tau,i})\{\widehat{m}_{2,t,h}(X_i,\tau)\}^2$$

at the estimated τ. The change-point parameter τ is estimated by minimization of the error of the model with a change-point at τ

$$\widehat{\tau}_{t,h} = \arg\min_{\tau \le t} \widehat{\sigma}^2_{\tau,t,h}$$

and the functions m_1 and m_2 by $\widehat{m}_{k,t,h}(x) = \widehat{m}_{k,t,h}(x, \widehat{\tau}_{t,h})$, for $k = 1, 2$. Let $\gamma = [T^{-1}\tau]$ and the corresponding change-point time $\tau_\gamma = T\gamma$, let

$m = (m_1, m_2)$ with true functions m_0, and let

$$\sigma_t^2(m, \gamma) = t^{-1} \sum_{i=1}^{t} \{Y_i - (I_{\tau_\gamma,i})m_1(X_i) - (1 - I_{\tau,i})m_2(X_i)\}^2$$

be the mean squared error for parameters (m, τ). The difference of the error from its minimal is

$$\begin{aligned} l_t(m,\tau) &= \sigma_t^2(m,\tau) - \sigma_t^2(m_0,\tau_0) \\ &= t^{-1}\sum_{i=1}^{t}\big[\{Y_i - I_{\tau,i}m_1(X_i) - (1-I_{\tau,i})m_2(X_i)\}^2 \\ &\qquad - \{Y_i - I_{\tau_0,i}m_{10}(X_i) - (1-I_{\tau_0,i})m_{20}(X_i)\}^2\big] \\ &= t^{-1}\sum_{i=1}^{t}\big[\{I_{\tau,i}m_1(X_i) - I_{\tau_0,i}m_{10}(X_i)\} \qquad (9.9) \\ &\qquad + \{(1-I_{\tau,i})m_2(X_i) - (1-I_{\tau_0,i})m_{20}(X_i)\}\big]^2, \\ &= t^{-1}\sum_{i=1}^{t}[\{(m_1-m_{10})(X_i)I_{\tau_0,i} - (m_2-m_{20})(X_i)(1-I_{\tau_0,i})\}^2 \\ &\qquad + \{(I_{\tau,i} - I_{\tau_0,i}))(m_1-m_2)(X_i)\}^2]\{1+o(1)\}. \end{aligned}$$

It converges *a.s.* to $l(m,\tau) = E_\alpha(m_1 - m_{10})^2(X) + E_\beta(m_2 - m_{20})^2(X) + |\tau - \tau_0|E(m_1 - m_2)^2(X)$ which is minimal for (m_0, τ_0), and the estimator $\widehat{\tau}_{nh}$ minimizes $l_t(\widehat{m}_{nh}, \tau)$. The *a.s.* consistency of the regression estimators $\widehat{m}_{nh} = (\widehat{m}_{1nh}, \widehat{m}_{2nh})$ and $l_t(m,\tau)$ imply that $\widehat{\gamma}_{t,h} = [t^{-1}\widehat{\tau}_{t,h}]$ is an *a.s.* consistent estimator of γ_0 in $]0,1[$. It follows that the estimator

$$\widehat{m}_{nh}(x) = \widehat{m}_{1nh}(x)I_{\widehat{\tau}_{t,h}} + \widehat{m}_{2nh}(x)(1 - I_{\widehat{\tau}_{t,h}})$$

of the regression function $m_0(x) = m_{10}(x)I_{\tau_0} + m_{20}(x)(1 - I_{\tau_0})$ is *a.s.* uniformly consistent and the process $(th)^{1/2}(\widehat{m}_{th} - m_0)$ converges weakly under P_{m_0} to a Gaussian process G_m on I_X, with mean and covariances zero and with variance function $V_m(x) = \kappa_2 Var(Y|X = x)$.

For the weak convergence of the change-point estimator, let $\|\varphi\|_X$ be the $L_2(F_X)$-norm of a function φ on I_X, $\rho(\theta,\theta') = (|\gamma - \gamma'| + \|m - m'\|_X^2)^{1/2}$ the distance between $\theta = (m^T, \gamma)^T$ and $\theta' = (m'^T, \gamma')^T$ and let $V_\varepsilon(\theta_0)$ be a neighbourhood of θ_0 with radius ε for the metric ρ. The quadratic function $l_t(m,\tau)$ defined by (9.9) converges to its expectation

$$l(\widehat{m}_{th}, \widehat{\tau}_{th}) = 0(\|\widehat{m}_{nh} - m_0\|_X^2 + |\widehat{\tau}_{nh} - \tau_0|).$$

The process is bounded in the same way

$$l_t(m,\tau) = [t^{-1}\sum_{i=1}^{t}\{(m_1 - m_{10})^2(X_i)I_{\tau_0,i} + (1 - I_{\tau_0,i})(m_2 - m_{20})^2(X_i)\}$$
$$+ t^{-1}\sum_{i=1}^{t}(I_{\tau,i} - I_{\tau_0,i})^2(m_1 - m_2)^2(X_i)]\{1 + o(1)\}.$$

it is denoted $l_t = (l_{1t} + l_{2t})\{1 + o(1)\}$. The process $W_t(m,\gamma) = t^{1/2}(l_t - l)(m,\tau_\gamma)$ is a $O_p(1)$. The estimator $\widehat{m}_{th}$ is a local maximum likelihood estimators of the nonparametric regression functions and the estimator of the change-point is a maximum likelihood estimator. The variable $l_{1t}(\widehat{m}_{th})$ converges to $l_1(m_0) = 0$ and $l_{2t}(\widehat{m}_{th}, \tau_{\widehat{\gamma}_t})$ converges to zero with the same rate if the convergence rate of $\widehat{\gamma}_t$ is the same as $\widehat{m}_{th}$. We obtain the next bounds.

Lemma 9.1. *For every $\varepsilon > 0$, there exists a constant κ_0 such that $E\sup_{(m,\gamma)\in V_\varepsilon(\tau_0)} l_t(m,\tau_\gamma) \le \kappa_0\varepsilon^2$ and $0 \le l(m,\tau_\gamma) \le \kappa_0\rho^2(\theta,\theta_0)$, for every θ in $V_\varepsilon(\tau_0)$.*

Lemma 9.2. *For every $\varepsilon > 0$, there exist a constant κ_1 such that $E\sup_{(m,\gamma)\in V_\varepsilon(\tau_0)} W_t(m,\gamma) \le \kappa_1\rho(\theta,\theta_0)$.*

The lemmas imply that for every $\varepsilon > 0$

$$\limsup_{t\to\infty, A\to\infty} P_0(th_t|\widehat{\gamma}_{th_t} - \gamma_0| > A) = 0.$$

The proof is similar to Ibragimov and Has'minskii's (1981) for a change-point of a density. It implies that $l_t(\widehat{\theta}_{th}) = (l_{1t} + l_{2t})(\widehat{\theta}_{th}) + o_p(1)$ uniformly. For the weak convergence of $(th)^{1/2}(\widehat{\gamma}_{th} - \gamma_0)$, let

$$\mathcal{U}_n = \{u = (u_m^T, u_\gamma)^T : u_m = (th)^{-1/2}(m - m_0), u_\gamma = (th)^{-1}(\gamma - \gamma_0)\}$$

be a bounded set. For every $A > 0$, let $\mathcal{U}_{th}^A = \{u \in \mathcal{U}_t; \|u\|^2 \le A\}$. Then for every $u = (u_m, u_\gamma)$ belonging to $\mathcal{U}_{th}^A$, $\theta_{t,u} = (m_{t,u}, \gamma_{t,u})$ with $m_{t,u} = m_0 + (th)^{-1/2}u_m$ and $\gamma_{t,u} = \gamma_0 + (th)^{-1}u_\gamma$. The process W_t defines a map $u \mapsto W_t(\theta_{t,u})$.

Theorem 9.1. *For every $A > 0$, the process $W_t(\theta)$ develops as*

$$W_t(\theta) = W_{1t}(m) + W_{2t}(\gamma) + o_p(1),$$

where the o_p is uniform on $\mathcal{U}_t^A$, as t tends to infinity. Then change-point estimator of γ_0 is asymptotically independent of the estimators of the regression functions $\widehat{m}_{1th}$ and $\widehat{m}_{2th}$.

Proof. For an ergodic process, the continuous part l_{1t} of l_t converges to

$$l_1(m) = \gamma_0 \|m_1 - m_{01}\|^2_{F_{1X}} + (1-\gamma_0)\|m_2 - m_{02}\|^2_{F_{2X}}$$

and the continuous part of W_t is approximated by $W_{1t}(m) = t^{1/2}(l_{1t} - l_1)(m)$. On $\mathcal{U}^A_{th}$, it is written

$$W_{1t}(m) = \{\gamma_0 \int (m_1 - m_{01})^2 \, d\nu_{1t} + (1-\gamma_0) \int (m_2 - m_{02})^2 \, d\nu_{2t}\},$$

where $\nu_{kt} = t^{1/2}(\widehat{F}_{kt} - F_{k0})$ is the empirical processes of the series in phase $k = 1, 2$, with the ergodic distributions F_{k0} of the process.

The discrete part of W_t is approximated by $W_{2t}(\gamma) = t^{1/2}(l_{2t} - l_2)(\gamma)$ where $l_{2t} = t^{-1}\sum_{i=1}^t (I_{\tau,i} - I_{\tau_0,i})^2(m_{10} - m_{20})^2(X_i) + o_p(|\tau - \tau_0|)$ and the sum is developed with the notation $a_i = (m_{10} - m_{20})^2(X_i)$

$$\begin{aligned} t^{-1}\sum_{i=1}^t (I_{\tau,i} - I_{\tau_0,i})^2(m_{10} - m_{20})^2(X_i) &= \frac{1}{t}\{\sum_{i=1}^{\tau_0}(1 - I_{\tau,i})a_i + \sum_{i=1+\tau_0}^{t} I_{\tau,i}a_i\} \\ &= \frac{1}{t}\{1_{\{\tau_{th}<\tau_0\}} \sum_{i=1+\tau_{th}}^{\tau_0} a_i + 1_{\{\tau_0<\tau_{th}\}} \sum_{i=1+\tau_0}^{\tau_{th}} a_i\} \\ &= \frac{1}{t}\{1_{\{\tau_{th}<\tau_0\}} \sum_{i=1+\tau_0-[h^{-1}u_\gamma]}^{\tau_0} a_i + 1_{\{\tau_0<\tau_{th}\}} \sum_{i=1+\tau_0}^{\tau_0+[h^{-1}u_\gamma]} a_i\}. \end{aligned}$$

Then l_{2t} converges uniformly to

$$l_2(\gamma) = |\gamma - \gamma_0|\{\int_X (m_{10} - m_{20})^2 \, dF_{2X} - \int_X (m_{10} - m_{20})^2 \, dF_{1X}\}$$

as h_t tends to zero. Let $\nu_{\tau,kt}$, $k = 1, 2$, be the empirical processes reduced to the variables between τ_0 and $\tau_0 + [h^{-1}u_\gamma]$, according to the sign of u_γ, and normalized by $|\tau - \tau_0|^{1/2}$. Then the process W_{2t} is approximated by

$$W_{2t}(\gamma) = |\gamma - \gamma_0|^{-1/2}\{\int_X (m_{10} - m_{20})^2 \, d\nu_{2t} - \int_X (m_{10} - m_{20})^2 \, d\nu_{\tau,1t}\}.$$

□

The limit of the process W_{1t} is a Gaussian distribution G_m. The estimator of the change-point satisfies

$$u_{\widehat{\gamma}_t} = th(\widehat{\gamma}_t - \gamma_0) = \arg\min_{\mathbb{R}} W_{2t}(u_\gamma) + o_p(1).$$

Its convergence rate is th and the asymptotic behaviour of the estimator $u_{\widehat{\gamma}_t}$ is deduced from Theorem 9.1 and the continuity of the minimum.

Theorem 9.2. *The change-point estimator* $th(\widehat{\gamma}_{th} - \gamma_0)$ *converges weakly to* $\arg\min_{u\in\mathbb{R}} Q(u)$ *and it is a* $O_p(1)$.

9.6 Exercises

(1) Let X_t be an AR(1) process $X_t = \alpha X_{t-1} + \varepsilon_t$, with an initial random variable X_0 with mean μ_0 and variance σ_0^2. Write it as a moving average and calculate the mean and the variance of the MA process.

(2) Consider the moving average process $X_t = \varepsilon_t - \theta_1 \varepsilon_{t-1} - \theta_2 \varepsilon_{t-2}$ with independent and identically distributed errors with mean zero and variance σ^2. Calculate the variance of X_t and the covariances between X_t and X_{t-k} and their empirical estimators. Extend to a model MA(q) of order q.

(3) Let $X_t = p_{k,t} + \theta \varepsilon_{t-1} + \sigma \varepsilon_t$, where $p_{k,t}$ is a polynomial of degree k. Define moving average estimators of $p_{k,t}$.

(4) Let X_t be an ARMA(2,2) process $X_t = \mu_t + \alpha X_{t-1} + \beta X_{t-2} + \varepsilon_t - \theta \varepsilon_{t-1}$, t in $\mathbb{N}$, with a constant or a varying mean and with noise variables with first moments $(0, \sigma^2)$. Define estimators for the parameters α, β and σ^2 and describe the series as infinite combinations of their past values.

(5) Invert the AR(1) model for X_t as a moving average model depending on X_0 and the noises and identify the parameters of both model in order to estimate them from Section 9.1.

(6) Define maximum likelihood estimators for the parameters α, β and σ^2 of the model $X_t = \mu_t(\theta) + \alpha X_{t-1} + \beta X_{t-2} + \sigma \varepsilon_t$, t in $\mathbb{N}$, with a parametric varying mean, independent identically distributed normal errors ε_i. For a polynomial trend $\mu_t(\theta)$, use differences of the series before the estimation of α and σ^2.

Chapter 10

Appendices

10.1 Appendix A

The moments of derivatives of the kernel estimator for the regression function are presented in Chapter 3, here the proofs are detailed. The variance of $\widehat{m}^{(1)}_{n,h}(x) = \widehat{f}^{-1}_{n,h}(x)\{\widehat{\mu}^{(1)}_{n,h}(x) - \widehat{m}_{n,h}(x)\widehat{f}^{(1)}_{n,h}(x)\}$ is obtained by an approximation similar to (3.2) in Proposition 3.1

$$\begin{aligned}
Var\widehat{m}^{(1)}_{n,h}(x) = f_X^{-2}[&Var\{\widehat{\mu}^{(1)}_{n,h}(x) - \widehat{m}_{n,h}(x)\widehat{f}^{(1)}_{n,h}(x)\} \\
&+ \{\mu^{(1)}_{n,h}(x) - m_{n,h}(x)f^{(1)}_{n,h}(x)\}^2 Var\widehat{f}_{n,h}(x) \\
&- 2\{\mu^{(1)}_{n,h}(x) - m_{n,h}(x)f^{(1)}_{n,h}(x)\} \\
&\times Cov\{\widehat{f}_{n,h}(x), \widehat{\mu}^{(1)}_{n,h}(x) - \widehat{m}_{n,h}(x)\widehat{f}^{(1)}_{n,h}(x)\}]\,\{1 + o(1)\},
\end{aligned}$$

where the variances $Var\widehat{\mu}^{(1)}_{n,h}(x)$ and $Var\widehat{f}^{(1)}_{n,h}(x)$ are $O((nh^3)^{-1})$, $Var\widehat{f}_{n,h}(x) = O((nh)^{-1})$, $E\{\widehat{f}^{(1)}_{n,h}(x) - f^{(1)}_{n,h}(x)\}^4 = O((nh^3)^{-1})$ and $E\{\widehat{m}_{n,h}(x) - mn, h(x)\}^4 = O((nh)^{-1})$,

$$\begin{aligned}
&Var\{\widehat{\mu}^{(1)}_{n,h}(x) - \widehat{m}_{n,h}(x)\widehat{f}^{(1)}_{n,h}(x)\} = Var\widehat{\mu}^{(1)}_{n,h}(x) \\
&+ Var\{\widehat{m}_{n,h}(x)\widehat{f}^{(1)}_{n,h}(x)\} - 2Cov\{\widehat{\mu}^{(1)}_{n,h}(x), \widehat{m}_{n,h}(x)\widehat{f}^{(1)}_{n,h}(x)\}, \\
&Var\{\widehat{m}_{n,h}(x)\widehat{f}^{(1)}_{n,h}(x)\} \le [E\{\widehat{m}_{n,h}(x)\}^4 E\{\widehat{f}^{(1)}_{n,h}(x)\}^4]^{1/2} \\
&\qquad - E^2\{\widehat{m}_{n,h}(x)\widehat{f}^{(1)}_{n,h}(x)\} = O((nh^2)^{-1})
\end{aligned}$$

Therefore $Var\{\widehat{\mu}^{(1)}_{n,h}(x) - \widehat{m}_{n,h}(x)\widehat{f}^{(1)}_{n,h}(x)\}$ and $Var\widehat{m}^{(1)}_{n,h}(x)$ are $O((nh^3)^{-1})$.

Proposition 10.1.

$$\begin{aligned}
Var\widehat{m}^{(1)}_{n,h}(x) &= f_X^{-2} Var\widehat{\mu}^{(1)}_{n,h}(x) + o((nh^3)^{-1}) \\
&= (nh^3)^{-1}\{f_X^{-2}(x)w_2(x)\int K^{(1)2} + o(h)\}.
\end{aligned}$$

10.2 Appendix B

In Chapter 4, the bandwidth is a real function defined on $\mathcal{I}_X$ and the normalized kernel is $\varphi_n(x) = K_{h_n(x)}(x)$. Its derivative with respect to x is

$$\varphi_n^{(1)}(x) = \frac{1}{h_n(x)} \frac{d}{dx} K\Big(\frac{x}{h_n(x)}\Big) - \frac{h_n^{(1)}(x)}{h_n^2(x)} K_{h_n(x)}(x)$$
$$= \frac{1}{h_n^2(x)} K^{(1)}\Big(\frac{x}{h_n(x)}\Big)\Big\{1 - x\frac{h_n^{(1)}(x)}{h_n(x)}\Big\} - \frac{h_n^{(1)}(x)}{h_n^2(x)} K_{h_n(x)}(x).$$

As $\|h_n\|$ tends to zero, a Taylor expansion of the density in a neighborhood of x and Lemma 2.1 yield

$$\int \varphi_n^{(1)}(x-u) f(u)\,du = f^{(1)}(x) - h_n(x) m_{2K}\Big\{\frac{h_n(x)}{2} f^{(3)}(x)$$
$$+ 2h_n^{(1)}(x) f^{(2)}(x) + o(\|h_n\| + \|h_n^{(1)}\|)\Big\}.$$

The expectation of the quadratic variations $|\widehat{f}_{n,h}(x) - \widehat{f}_{n,h}(y)|^2$ develops as the sum

$$n^{-1} \int \{K_{h_n(x)}(x-u) - K_{h_n(y)}(y-u)\}^2 f(u)\,du$$
$$+ (1-n^{-1})\{f_{n,h_n}(x) - f_{n,h_n}(y)\}^2.$$

For an approximation of the first term, the Mean Value Theorem implies $K_{h_n(x)}(x-u) - K_{h_n(y)}(y-u) = (x-y)\varphi_n^{(1)}(z-u)$ where z is between x and y, then $\int \{K_{h_n(x)}(x-u) - K_{h_n(y)}(y-u)\}^2 f(u)\,du$ is approximated by

$$(x-y)^2 \int \varphi_n^{(1)2}(z-u) f(u)\,du = (x-y)^2 h_n^{-3}(x)\Big\{f(x)\int K^{(1)2} + o(\|h_n\|)\Big\}.$$

Since $h_n^{-1}(x)|x|$ and $h_n^{-1}(y)|y|$ are bounded by 1, the order of $E|\widehat{f}_{n,h}(x) - \widehat{f}_{n,h}(y)|^2 = O((x-y)^2 n^{-1}\|h_n^{-1}\|$ if $|xh_n(y) - yh_n(x)| \le 2h_n(y)h_n(x)$ and it is a sum of variances otherwise.

10.3 Appendix C

In the single index model studied in Chapter 7, the precise order of the mean and variance of $\widehat{V}_{n,h}^{(1)}$ defined Section 7.2 requires expansions. The empirical mean squared error of the estimated function g, at fixed θ has the derivative

$$\widehat{V}_{n,h}^{(1)}(\theta) = n^{-1} \sum_{i=1}^n \{Y_i - \widehat{g}_{n,h}(\theta^T X_i;\theta)\}\,\{\widehat{g}_{n,h}^{(1)}(\theta^T X_i;\theta)\}\,X_i.$$

Let $Z = \theta^T X$ at fixed θ. The mean of $\widehat{V}_{n,h}^{(1)}$ is

$$\begin{aligned} V_{n,h}^{(1)}(\theta) &= EE[\{g(Z_i) - \widehat{g}_{n,h}(Z_i;\theta)\}\,\{\widehat{g}_{n,h}^{(1)}(Z_i;\theta)\}\,X_i \mid X_i] \\ &= E[\{b_{g^{(1)},n,h} g_{n,h}^{(1)}(Z_i;\theta) - Cov(\widehat{g}_{n,h}, \widehat{g}_{n,h}^{(1)})(Z_i;\theta)\}\,X_i^{\otimes 2}]. \end{aligned}$$

Its variance is $n^{-1} Var[\{Y_i - \widehat{g}_{n,h}(Z_i;\theta)\}\,\{\widehat{g}_{n,h}^{(1)}(Z_i;\theta)\}\,X_i]$ and,

$$\begin{aligned} Var[\{Y_i - \widehat{g}_{n,h}(Z_i;\theta)\}\,\widehat{g}_{n,h}^{(1)}(Z_i;\theta)\} &= O(Var\{Y_i\,\widehat{g}_{n,h}^{(1)}(Z_i;\theta)\}) \\ &\quad + O(Var\{\widehat{g}_{n,h}(Z_i;\theta)\,\widehat{g}_{n,h}^{(1)}(Z_i;\theta)\}) \\ Var\{Y_i\,\widehat{g}_{n,h}^{(1)}(Z_i;\theta)\} &= O(Var\{\widehat{g}_{n,h}^{(1)}(Z_i;\theta)\}) = O((nh^3)^{-1}). \end{aligned}$$

The expansions (3.2) for $\widehat{g}_{n,h}$ and (3.15) for $\widehat{g}_{n,h}^{(1)}$ are written

$$\begin{aligned} \{\widehat{g}_{n,h} - g_{n,h}\}(z) &= f_X^{-1}(z)\big\{(\widehat{\mu}_{n,h} - \mu_{n,h})(z) \\ &\quad - g(z)(\widehat{f}_{X,n,h} - f_{X,n,h})(z)\big\} + o_{L_2}((nh)^{-1/2}) \\ \{\widehat{g}_{n,h}^{(1)} - g_{n,h}^{(1)}\}(z) &= f_X^{-1}(z)[(\widehat{\mu}_{n,h}^{(1)} - \mu_{n,h}^{(1)})(z) \\ &\quad - \{\widehat{g}_{n,h}\widehat{f}_{X,n,h}^{(1)} - E(\widehat{g}_{n,h}\widehat{f}_{X,n,h}^{(1)})\}(z) \\ &\quad - g^{(1)}(z)\,(\widehat{f}_{X,n,h} - f_{X,n,h})(z)] + o_{L_2}((nh^3)^{-1/2}), \end{aligned}$$

$$\begin{aligned} \widehat{g}_{n,h}(z)\widehat{f}_{X,n,h}^{(1)}(z) &= \{\widehat{f}_{X,n,h}^{(1)} - f_{X,n,h}^{(1)}\}(z)[g_{n,h} + f_X^{-1}\big\{(\widehat{\mu}_{n,h} - \mu_{n,h}) \\ &\quad - g(\widehat{f}_{X,n,h} - f_{X,n,h})\big\}](z) \\ &\quad + f_{X,n,h}^{(1)}(z)[g_{n,h} + f_X^{-1}\big\{(\widehat{\mu}_{n,h} - \mu_{n,h}) \\ &\quad - g(\widehat{f}_{X,n,h} - f_{X,n,h})\big\}](z) + o_{L_2}((nh)^{-1/2}) \\ E\{\widehat{g}_{n,h}\widehat{f}_{X,n,h}^{(1)}\}(z) &= f_X^{-1}(z)\,E\{\widehat{f}_{X,n,h}^{(1)} - f_{X,n,h}^{(1)}\}(z)\big\{(\widehat{\mu}_{n,h} - \mu_{n,h}) \\ &\quad - g(\widehat{f}_{X,n,h} - f_{X,n,h})\big\}(z) + f_{X,n,h}^{(1)}(z)g_{n,h}(z) \\ &\quad + o_{L_2}(nh^2) = f_{X,n,h}^{(1)}(z)g_{n,h}(z) + O((nh^2)^{-1}). \end{aligned}$$

Then

$$\begin{aligned} &\{\widehat{g}_{n,h}\widehat{f}_{X,n,h}^{(1)} - E(\widehat{g}_{n,h}\widehat{f}_{X,n,h}^{(1)})\}(z) = \{\widehat{f}_{X,n,h}^{(1)} - f_{X,n,h}^{(1)}\}(z)([g_{n,h} \\ &+ f_X^{-1}\big\{(\widehat{\mu}_{n,h} - \mu_{n,h}) - g(\widehat{f}_{X,n,h} - f_{X,n,h})\big\}](z) \\ &+ f_{X,n,h}^{(1)}(z)f_X^{-1}\big\{(\widehat{\mu}_{n,h} - \mu_{n,h}) - g(\widehat{f}_{X,n,h} - f_{X,n,h})\big\}(z)) + o_{L_2}((nh)^{-1/2}) \end{aligned}$$

$$\begin{aligned}
\{\widehat{g}_{n,h}^{(1)} - g_{n,h}^{(1)}\}(z) &= f_X^{-1}(z)[(\widehat{\mu}_{n,h}^{(1)} - \mu_{n,h}^{(1)})(z) - (\widehat{f}_{X,n,h}^{(1)} - f_{X,n,h}^{(1)})(z) \\
&\quad \times ([g_{n,h} + f_X^{-1}\{(\widehat{\mu}_{n,h} - \mu_{n,h}) - g(\widehat{f}_{X,n,h} - f_{X,n,h})\}](z) \\
&\quad + [f_{X,n,h}^{(1)} f_X^{-1}\{(\widehat{\mu}_{n,h} - \mu_{n,h}) - g(\widehat{f}_{X,n,h} - f_{X,n,h})\}](z)) \\
&\quad - g^{(1)}(z)\,(\widehat{f}_{X,n,h} - f_{X,n,h})(z)] + o_{L_2}((nh)^{-1/2}) \\
Cov(\widehat{g}_{n,h}, \widehat{g}_{n,h}^{(1)})(z) &= -f_X^{-2}(z)[Cov(\widehat{\mu}_{n,h}, \widehat{\mu}_{n,h}^{(1)}) - g^{(1)} Cov(\widehat{\mu}_{n,h}, \widehat{f}_{X,n,h}) \\
&\quad - g Cov(\widehat{f}_{X,n,h}, \widehat{\mu}_{n,h}^{(1)}) - g g^{(1)} Var \widehat{f}_{X,n,h} - g_{n,h} \\
&\quad - g_{nh} Cov(\widehat{f}_{X,n,h}, \widehat{\mu}_{n,h}^{(1)}) + g g_{nh} Cov(\widehat{f}_{X,n,h}, \widehat{f}_{n,h}^{(1)}) \\
&\quad + f_{n,h}^{(1)} Var \widehat{g}_{n,h} + E\{(\widehat{f}_{n,h}^{(1)} - f_{n,h}^{(1)})(\widehat{g}_{n,h} - g_{n,h}] \\
&\quad + o((nh^2)^{-1})
\end{aligned}$$

where the main term has the same order as $Cov(\widehat{\mu}_{n,h}, \widehat{\mu}_{n,h}^{(1)})$, namely $O((nh^2)^{-1})$ and the variances and covariances for the terms without derivatives are $O((nh)^{-1})$. The product and $E\{\widehat{g}_{n,h}\widehat{g}_{n,h}^{(1)}\}(z) = g_{n,h}(z)g_{n,h}^{(1)}(z) + O((nh^2)^{-1})$, then

$$\begin{aligned}
Var\{\widehat{g}_{n,h}\widehat{g}_{n,h}^{(1)}\}(z) &= E[\{\widehat{g}_{n,h}(z) - g_{n,h}(z)\}^2\{\widehat{g}_{n,h}^{(1)}(z) - g_{n,h}^{(1)}(z)\}^2 \\
&\quad + g_{n,h}^2(z) Var\{\widehat{g}_{n,h}^{(1)}(z)\} + g_{n,h}^{(1)2}(z) Var\{\widehat{g}_{n,h}(z)\} \\
&\le E[\{\widehat{g}_{n,h}(z) - g_{n,h}(z)\}^4]^{1/2} E[\{\widehat{g}_{n,h}^{(1)}(z) - g_{n,h}^{(1)}(z)\}^4]^{1/2} \\
&\quad + g_{n,h}^2(z) Var\{\widehat{g}_{n,h}^{(1)}(z)\} + g_{n,h}^{(1)2}(z) Var\{\widehat{g}_{n,h}(z)\}.
\end{aligned}$$

Finally,

$$\begin{aligned}
E[\{\widehat{g}_{n,h}(z) - g_{n,h}(z)\}^4] &= O((nh)^{-1}), \\
E[\{\widehat{g}_{n,h}^{(1)}(z) - g_{n,h}^{(1)}(z)\}^4] &= O((nh^3)^{-1}), \\
Var\{\widehat{g}_{n,h}\widehat{g}_{n,h}^{(1)}\}(z) &= O((nh^3)^{-1})
\end{aligned}$$

therefore the main term of the variance of the product $\widehat{g}_{n,h}(z)\widehat{g}_{n,h}^{(1)}(z)$ is $g_{n,h}^2(z) Var\{\widehat{g}_{n,h}^{(1)}(z)\} = O((nh^3)^{-1})$ and the mean $V_{n,h}^{(1)}(\theta)$ is a $O(h^2) + O((nh^2)^{-1})$.

Lemma 10.1. *For* $1 \le i \ne j \le n$

$$E\{\varphi(X_i) - \varphi(X_j)\}K_h(X_i - X_j) = \frac{h^2 m_{2K}}{2} E(f_X \varphi^{(2)} + f_X^{(1)} \varphi^{(1)})(X) + r_{1n},$$

$$E\{\varphi(X_i) - \varphi(X_j)\}K_h^2(X_i - X_j) = E[\{f_X \varphi^{(1)}\}(X)] \int z K^2(z)\, dz + r_{2n},$$

where $r_{1n} = o(h^2)$ *and*

$$r_{2n} = h \int z^2 K^2(z)\, dz E[\{f_X \varphi^{(2)} + 2\varphi^{(1)} f_X^{(1)}\}(X)] + o(h^2).$$

Proof. Using Lemma 2.1 and an expansion for $|X_i - X_j| \le h$

$$\begin{aligned} &E\{\varphi(X_i) - \varphi(X_j)\}K_h^2(X_i - X_j) \\ &= \int\int \{\varphi(x) - \varphi(y)\}K_h^2(x-y)f_X(x)f_X(y)\,dx\,dy \\ &= h^{-1}\int\int \{hz\varphi^{(1)} + \frac{z^2h^2}{2}\varphi^{(2)} + o(h)\}\{f_X + hzf_X^{(1)} + o(h^2)\} \\ &\qquad\qquad \times K^2(z)\,dF_X\,dz. \end{aligned}$$

The higher orders are obtained by further terms in the expansion. □

10.4 Appendix D

The ergodicity and mixing conditions for the convergence of functionals of a process $(X_t)_{t\ge 0}$ or a sequence of dependent variables $(X_i)_{1\le i\le n}$ are expressed in following conditions A and B. Let $(X_i)_{1\le i\in\mathbb{N}}$ be a sequence of dependent random variable with values in a metric space $(\mathbb{X}, \mathcal{A}, \mu)$ and such that EX_i^2 is finite for every i. Let $\mathcal{M}_1^j$ and $\mathcal{M}_{j+k}^\infty$ be the σ-algebras generated by the sub-samples $(X_i)_{1\le i\le j}$ and, respectively, $(X_i)_{j+k\le i}$. The sequence $(X_i)_{1\le i}$ is φ-mixing if there exists a real sequence $(\varphi_k)_{k\in\mathbb{N}}$ converging to zero as k tends to infinity and such that the conditional probabilities of the sample $(X_i)_{i\in\mathbb{N}}$ satisfy

$$\sup\{|P(B|A) - P(B)|; A \in \mathcal{M}_1^j, B \in \mathcal{M}_{j+k}^\infty, j, k \in \mathbb{N}\} < \varphi_k.$$

A1 The sequence $(X_i)_{i\ge 1}$ is ergodic if there exists a probability ν on $(\mathbb{X}, \mathcal{A}, \mu)$ such that for every real bounded function ϕ on $\mathbb{X}$

$$\frac{1}{n}\sum_{i=1}^n \phi(X_i) \xrightarrow[n\to\infty]{P} E_\nu\phi(X_1).$$

B1 The sequence $(X_i)_{i\ge 0}$ is φ-mixing with a sequence $(\varphi_k)_{k\ge 1}$ satisfying $\sum_{k\ge 1}(k+1)^2\varphi_k^{1/2} < \infty$.

The φ-mixing property and condition B1 are defined in Billingsley (1968), they imply the weak convergence of the normalized variable $n^{1/2}\{\frac{1}{n}\sum_{i=1}^n \phi(X_i) - E_\nu\phi(X_1)\}$ to a normal variable $\sigma_\varphi^2\mathcal{N}(0,1)$.

Let $(X_t)_{t\ge 0}$ be a time indexed process such that for every $t > 0$, X_t is a random variable in a metric space $(\mathbb{X}, \mathcal{A}, \mu)$ and EX_t^2 is finite. Let $\mathcal{M}_0^s$ and $\mathcal{M}_{s+t}^\infty$ be the σ-algebras generated by the sample-paths of the process observed on the time intervals $[0, s]$ and $[s+t, \infty[$ respectively,

$\mathcal{M}_0^s = \sigma\{(X_u)_{0\le u\le s}\}$ and $\mathcal{M}_{s+t}^\infty = \sigma\{(X_u)_{u\ge s+t}\}$, with s and $t > 0$. The sequence $(X_t)_{t\ge0}$ is φ-mixing if there exists a sequence $(\varphi_t)_{t\ge0}$ converging to zero as t tends to infinity and such that the marginal distributions of the process $(X_t)_{t\ge0}$ satisfy

$$\sup\{|P(B|A) - P(B)|; A \in \mathcal{M}_0^s, B \in \mathcal{M}_{s+t}^\infty, s, t \ge 0\} < \varphi_t.$$

The ergodicity and mixing conditions are modified as follows.

A2 The process $(X_t)_{t\ge0}$ is ergodic if there exists a probability ν on $(\mathbb{X}, \mathcal{A}, \mu)$ such that for every real bounded function ϕ defined on the space $\mathbb{X}$

$$\frac{1}{t}\int_0^t \phi(X_s)\,ds \xrightarrow[t\to\infty]{P} \int_{\mathbb{X}} \phi(x)d\nu(x).$$

B2 The process $(X_t)_{t\ge0}$ is φ-mixing with a sequence $(\varphi_t)_{t\ge0}$ satisfying $\int_0^\infty (t+1)^2 \varphi_t^{1/2}\,dt < \infty$.

The ergodic property is strengthened to allow the convergence of functionals of the joint distributions of the process at several observation times.

A2' The process $(X_t)_{t\ge0}$ is ergodic if for every integer k there exists a probability ν_k on $(\mathbb{X}^{\otimes k}, \mathcal{A}_k, \mu)$, with the Borel σ-algebra $\mathcal{A}_k$ on $\mathbb{X}^{\otimes k}$, such that for every real bounded function ϕ defined on the space $\mathbb{X}^{\otimes k}$

$$\frac{1}{t^k}\int_{[0,t]^k} \phi(X_{s_1}, \ldots, X_{s_k})\,ds_1, \ldots, ds_k \xrightarrow[t\to\infty]{P} E_{\nu_k}\phi(X_1, \ldots, X_k).$$

The expectation with respect to the limit ν_k is an integral on $\mathbb{X}^{\otimes k}$, $E_{\nu_k}\phi(X_1, \ldots, X_k) = \int_{\mathbb{X}^{\otimes k}} \phi(x_1, \ldots, x_k)\,\nu_k(dx_1, \ldots, dx_k)$. For every integer $k > 0$, this property is the consistency of the k-th moment of the process X. For $k = 2$, it implies the convergence in probability of the covariance function of the process X. Condition B2 entails the weak convergence of the process $t^{1/2}\{t^{-1}\int_0^t \phi(X_s)\,ds - E_\nu\phi(X_1)\}$ to a normal process $\sigma_\varphi^2 W_1$. The φ-mixing property A1 and condition B1 imply the consistency and the weak convergence of the partial sum of the sequence of variables $Z_k = X(n^{-1}k) - X(n^{-1}(k-1))$, $1 \le k \le n$. Under condition A1, the process $S_n(x) = n^{-1/2}\sum_{k=1}^{[nx]} Z_k$, x in $[0,1]$, converges weakly to the Brownian motion defined on $[0,1]$ and with covariance function $C(s,t) = s \wedge t$.

Notations

1_A	indicator of a set A,
$a.s.$	almost surely,
$(B_t)_{t\geq 0}$	Brownian motion,
$Cov(X_i, X_j)$	covariance of X_i and X_j: $E(X_i - EX_i)(X_j - EX_j)$,
$C_s(I)$	class of real functions on I, having bounded and continuous derivatives of order s,
$\Delta f(x, y)$	variation of f: $f(y) - f(x)$,
EX	expectation (or mean) of a variable X: $\int x\, dF_X(x)$,
$F_X(x)$	probability of the random set $\{X \leq x\}$,
$f^{(s)}$	derivatives of order s for a function f,
$\widehat{F}_{X,n}$	empirical distribution function,
m^{-1}	either $1/m$ or inverse of a monotone function m,
$\mathcal{H}_{\alpha,M}$	class of real functions f such that $\forall x$ and y, $\lvert f^{(s)}(x) - f^{(s)}(y)\rvert \leq M\lvert x - y\rvert^{\alpha - s}$, $s = [\alpha]$,
$K_h = h^{-1}K(h^{-1}\cdot)$	normalized kernel with bandwidth h,
$\Lambda = \int_0^{\cdot} \frac{dF}{(1-F^-)}$	cumulative hazard function for the distribution function F,
$N = (N_t)_{t\geq 0}$	point process,
$\widehat{\nu}_n$	empirical process $n^{1/2}(\widehat{F}_{X,n} - F_X)$,
$\mathcal{L}_n$	partial likelihood of N at t such that $N_t = n$,
$(\widetilde{N}_t)_{t\geq 0}$	predictable compensator of a point process N,
Ω	sample space,
$\rho(i, j)$	correlation of variables X_i and X_j: $Cov(X_i, X_j)\{VarX_i VarX_j\}^{-1/2}$,
$VarX$	variance of a variable X: $E(X - EX)^2$,
$\widehat{V}_{n,h}$	empirical mean squared error for a regression,

$(X_t)_{t\geq 0}$	continuously observed process or time series,
$(W_t)_{t\geq 0}$	Gaussian process,
$\mathcal{Z}_h$	$\{z \in \mathcal{Z}; \sup_{z' \in \partial\mathcal{Z}} \|z - z'\| \geq h\}$, with the frontier $\partial\mathcal{Z}$ of $\mathcal{Z}$.

Bibliography

Andersen, P. and Gill, R. D. (1982). Cox's regression model for counting processes: a large sample study, *Ann. Statist.* **10**, pp. 1100–1120.

Bahadur, R. R. (1966). A note on quantiles in large samples, *Ann. Math. Statist.* **37**, pp. 577–580.

Barlow, R., Bartholomew, D. J., Bremmer, J. and Brunk, H. D. (1972). *Statistical Inference under Order Restrictions* (Wiley, New York).

Beran, R. J. (1972). Upper and lower risks and minimax procedures, *Proceedings of the sixth Berkeley Symposium on Mathematical Statistics, L. Lecam, J. Neyman and E. Scott (eds)* , pp. 1–16.

Bickel, P. and Rosenblatt, P. (1973). On some global measures of the deviations of density functions estimates, *Ann. Statist.* **1**, pp. 1071–1095.

Billingsley, P. (1968). *Convergence of probability measures* (Wiley, New York).

Bosq, D. (1998). *Nonparametric Statistics for Stochastic Processes* (2nd edition, Springer, New York).

Bowman, A. W. (1984). An alternative method of cross-validation for the smoothing of density estimates, *Biometrika* **71**, pp. 353–360.

Bowman, A. W. and Azalini, A. (1997). *Applied Smoothing Techniques for Data Analysis. The Kernel Approach with S-Plus Illustrations* (Oxford Statistical Science Series 18).

Breslow, N. and Crowley, J. (1974). A large sample study of the life table and product limit estimates under random censorship, *Ann. Statist.* **2**, pp. 437–453.

Bretagnolle, J. and Huber, C. (1981). Estimation de densités : risque minimax, *Z. Wahrsch. Verw. Geb.* **47**, pp. 119–139.

Brillinger, D. R. (1981). *Time Series Data Analysis and Theory* (Holt, Rinehart and Winston, New York).

Chaudhuri, P. (1991). Nonparametric estimates of regression quantiles and their local Bahadur representation, *Ann. Statist.* **19**, pp. 760–777.

Chernoff, H. (1964). Estimation of the mode, *Ann. Inst. Statist. Math.* **16**, pp. 31–41.

Cox, D. R. and Oakes, D. (1984). *Analysis of Survival Data* (Chapman and Hall, London).

Cox, R. D. (1972). Regression model and life tables, *J. Roy. Statist. Soc. Ser. B* **34**, pp. 187–220.

De Boor, C. (1978). *A Practical Guide to Splines* (Springer, New York).

Deheuvels, P. (1977). Estimation non paramétrique de la densité par histogrammes généralisés, *Rev. Statist. Appl* **25**, pp. 5–42.

Delecroix, M., Härdle, W. and Hristache, M. (2003). Optimal smoothing in single-index models, *J. Multiv. Anal.* **286**, pp. 213–226.

Devroye, L. (1983). The equivalence of weak, strong and complete convergence in l1 for kernel density estimates, *Ann. Statist.* **11**, pp. 896–904.

Dumbgen, L. and Rufibach, K. (2009). Maximum likelihood estimatio of a log-concave density and its distribution function: Basic properties and uniform consistency, *Bernoulli* **15**, pp. 40–68.

Dvoretski, A., Kiefer, J. and Wolfowitz, J. (1956). Asymptotic minimax character of the sample distribution functions and of the classical multinomial estimator, *Ann. Math. Statist.* **27**, pp. 642–669.

Eubank, R. (1977). *Spline Smoothing and Nonparametric Regression* (Dekker, New York).

Fan, J. and Gijbels, I. (1996). *Polynomial Modelling and Its Applications* (Chapman and Hall CRC).

Ghosh, J. K. (1966). A new proof of the Bahadur representation of quantiles and an application, *Ann. Math. Statist.* **42**, pp. 1957–1961.

Gijbels, I. and Veraverbeke, N. (1988a). Almost sure asymptotic representation for a class of functionals of the product-limit estimator, *Ann. Statist.* **19**, pp. 1457–1470.

Gijbels, I. and Veraverbeke, N. (1988b). Weak asymptotic representations for quantiles of the product-limit estimator, *J. Statist. Plann. Inf.* **18**, pp. 151–160.

Groeneboom, P. (1989). Brownian motion with a parabolic dridt and airy functions, *Probab. Theory Related Fields* **81**, pp. 79–109.

Groeneboom, P., Jonkbloed, G. and Wellner, J. (2001). Estimation of a convex function: Characterization and asymptotic theory, *Ann. Statist.* **29**, pp. 1653–1698.

Groeneboom, P. and Wellner, J. (1990). *Empirical processes* (Burckhlder, Basel).

Guyon, X. and Perrin, O. (2000). Identification of space deformation using linear and superficial quadratic variations, *Statist. Prob. Lett.* **47**, pp. 307–316.

Hall, P. (1981). Law of the iterated logarithm for nonparametric density estimators, *Stoch. Proc. Appl.* **56**, pp. 47–61.

Hall, P. (1984). Integrated square error properties of kernel estimators of regression functions, *Ann. Statist.* **12**, pp. 241–260.

Hall, P. and Huang, L.-S. (2001). Nonparametric kernel regression subjet to monotonicity constraints, *Ann. Statist.* **29**, pp. 624–647.

Hall, P. and Johnstone, I. (1992). Empirical functionals and efficient smoothing parameter selection, *J. Roy. Statist. Soc. Ser. B* **54**, pp. 475–530.

Hall, P. and Marron, J. M. (1987). Estimation of integrated squared density derivatives, *Statist. Probab. Lett.* **6**, pp. 109–115.

Härdle, W. (1990). *Applied Nonparametric Regression* (Cambridge University Press).

Härdle, W. (1991). *Smoothing Methods in Statistics* (Cambridge University Press, UK).

Härdle, W., Hall, P. and Ihimura, H. (1993). Optimal smoothing in single-index models, *Ann. Statist.* **21**, pp. 157–178.

Härdle, W., Hall, P. and Marron, J. M. (1988). How far are asymptotically choosen regression smoothers from their optimum ? (with discussion), *J. Amer. Statist. Soc.* **6**, pp. 109–115.

Hirstache, M., Juditsky, A. and Spokoiny, V. (2001). Direct estimation of the index coefficients in a single-index model, *Ann. Statist.* **29**, pp. 595–623.

Ibragimov, I. and Has'minskii, R. (1981). *Statistical Estimation: Asymptotic Theory* (Springer, New York).

Ichimura, H. (1993). Semi-parametric least squares and weighted sls estimation of single-index models, *J. Econometrics* **58**, pp. 71–120.

Jones, M. C., Marron, J. S. and Park, B. U. (1991). A simple root n bandwidth selector, *Ann. Statist.* **19**, pp. 1919–1932.

Kaplan, M. and Meier, P. A. (1958). Nonparametric estimator from incomplete observations, *J. Am. Statist. Ass.* **53**, pp. 457–481.

Khasminskii, R. Z. (1992). *Topics in nonparametric estimation* (American Mathematical Society).

Kiefer, J. (1972). Iterated logaritm analogues for samples quantiles when $p_n \downarrow 0$, *Proceedings of the sixth Berkeley Symposium on Mathematical Statistics, L. Lecam, J. Neyman and E. Scott (eds)* , pp. 227–244.

Kiefer, J. and Wolfowitz, J. (1976). Asymptotically minimax estimation of concave and convex distribution functions, *Z. Wahrsch. Verw. Gebiete* **34**, pp. 73–85.

Kim, J. and Pollard, D. (1990). Cube root asymptotics, *Ann. Statist.* **18**, pp. 191–219.

Lecam, L. (1990). *On the asymptotic theory of estimation* (Springer, New York).

Lo, S.-H. and Singh, K. (1986). The product-limit estimator and the bootstrap: some asymptotic representations, *Prob. Theor. Rel. Fields* **71**, pp. 455–465.

Mammen, E. (1991). Estimating a smooth monotone regression function, *Ann. Statist.* **19**, pp. 724–740.

Marron, J. S. (1988). Improvment of a data based bandwidth selector, *Preprint Univ. North Carolina, Chapel Hill* , pp. 1–31.

Messer, K. (1991). A comparison of a spline estimate to its equivalent kernel estimate, *Ann. Statist.* **19**, pp. 817–829.

Meyer, M. and Woodroof, M. (2000). On the degrees of freedom in shaped-restricted regression, *Ann. Statist.* **28**, pp. 1083–1104.

Nadaraya, E. A. (1964). On estimating regression, *Theor. Probab. Appl.* **9**, pp. 141–142.

Nadaraya, E. A. (1989). *Nonparametric estimation of probability densities and regression curves* (Kluwer Academic Publisher, Boston).

Ould Saïd, E. (1997). A note on ergodic processes prediction via estimation of the conditional mode, *S. J. Statist.* **24**, pp. 231–239.

Parzen, E. A. (1962). On the estimation of probability density and mode, *Ann. Math. Statist.* **33**, pp. 1065–1076.

Parzen, E. A. (1979). Nonparametric statistical data modeling, *J. Amer. Statist. Assoc.* **74**, pp. 105–131.

Perrin, O. (1999). Quadratic variations for gaussian processes and application to time deformation, *Stoch. Proc. Appl.* **82**, pp. 293–305.

Pinçon, C. and Pons, O. (2006). Nonparametric estimator of a quantile function for the probability of event with repeated data, *Dependence in Probability and Statistics, Lect. N. Statist.* **17.** Springer, New York , pp. 475–489.

Plancherel, M. and Rotach, W. (1929). Sur les valeurs asymptotiques des polynômes d'Hermite, *Commentarii Mathem. Helvet.* **1**, pp. 227–254.

Pons, O. (1986). Vitesse de convergence des estimateurs à noyau pour l'intensité d'un processus ponctuel, *Statistics* **17**, pp. 577–584.

Pons, O. (2000). Nonparametric estimation in a varying-coefficient Cox model, *Mathematical Methods of Statistics* **9**, pp. 376–398.

Pons, O. (2007a). Estimation of absolutely continuous distributions for censored variables in two-samples nonparametric and semi-parametric models, *Bernoulli* **13**, pp. 92–114.

Pons, O. (2007b). Estimation of the distribution function of one and two dimensional censored variables or sojourn times of markov renewal processes, *Comm. Statist.–Theory Methods* **55**, pp. 1–18.

Pons, O. (2009a). *Estimation and tests in distribution mixtures and change-points models* (O. Pons, Viroflay, F).

Pons, O. (2009b). *Nonparametric Estimation for Renewal and Markov Processes* (O. Pons, Viroflay, F).

Pons, O. and de Turckheim, E. (1987). Estimation in Cox's periodic model with a histogram-type estimator for the underlying intensity, *Scand. J. Statist.* **14**, pp. 329–345.

Prakasa Rao, B. L. S. (1983). *Nonparametric Functional Estimation* (Academic Press, New York).

Rebolledo, R. (1978). Sur les applications de la théorie des martingales à l'étude statistique d'une famille de processus ponctuels, *Journée de Statistique des Processus Stochastiques, Lecture Notes in Mathematics* **636**, pp. 27–70.

Rice, J. and Rosenblatt, M. (1983). Smoothing splines: regression, derivatives and deconvolution, *Ann. Statist.* **11**, pp. 141–156.

Robinson, P. M. (1991). Automatic frequency domain inference on semiparametric and nonparametric models, *Econometrika* **59**, pp. 1329–1363.

Rosenblatt, M. (1956). Remarks on some nonparametric estimates of a density function, *Ann. Math. Statist.* **27**, pp. 832–837.

Rosenblatt, M. (1975). A quadratic measures of deviation of two-dimensional density estimates, *Ann. Statist.* **3**, pp. 1–14.

Rudemo, M. (1982). Empirical choice of histograms and kernel density estimators, *Scand. J. Statist.* **9**, pp. 65–78.

Ruppert, D., Wand, M. P. and Carroll, R. J. (2003). *Semiparametric Regression* (Cambridge University Press, UK).

Schoenberg, I. (1964). Spline functions and the problem of graduation, *Proc. Nat. Acad. Sci. USA* **52**, pp. 947–950.

Schuster, E. F. (1969). Estimation of a probability density function and its derivatives, *Ann. Math. Statist.* **40**, pp. 1187–1195.

Scott, D. W. (1992). *Multivariate density estimation: theory, practice, and visualization* (Wiley, New York).

Sheather, S. J. and Marron, J. S. (1990). Kernel quantile estimators, *J. Amer. Statist. Assoc.* **85**, pp. 410–416.

Shorack, G. R. and Wellner, J. A. (1986). *Empirical processes and applications to statistics* (Wiley, New York).

Silverman, B. W. (1978a). On a gaussian process related to multivariate probability density estimation, *Math. Proc. Cambridge Philos. Soc.* **80**, pp. 136–144.

Silverman, B. W. (1978b). Weak and strong uniform consistency of the kernel estimate of a density and its derivatives, *Ann. Statist.* **6**, pp. 17–184.

Silverman, B. W. (1984). Spline smoothing: The equivalent variable kernel method, *Ann. Statist.* **12**, pp. 898–916.

Silverman, B. W. (1985). Some aspects of the spline smoothing approach to the nonparametric regression curve fitting, *J. Roy. Statist. Soc. Ser. B* **47**, pp. 1–22.

Simonoff, J. S. (1996). *Smoothing Methods in Statistics* (Springer-Verlag, New York).

Singh, R. S. (1979). On necessary and sufficient conditions for uniform strong consistency of estimators of a density and its derivatives, *J. Multiv. Anal.* **9**, pp. 157–164.

Stieltjes, T.-J. (1890). Sur les polynômes de Legendre, *Ann. Fac. Sci. Toulouse, 1e série* **4 G**, pp. 1–17.

Stone, M. (1974). Cross-validation choice and assessment of statistical prediction (with discussion), *J. Roy. Statist. Soc. Ser. B* **36**, pp. 111–147.

Stute, W. (1982). A law of the logarithm for kernel density estimators, *Ann. Probab.* **10**, pp. 414–422.

van de Geer, S. (1993). Hellinger consistency of certain nonparametric maximum likelihood estimators, *Ann. Statist.* **21**, pp. 14–44.

van de Geer, S. (1996). *Applications of empirical process theory* (Cambridge university press).

van der Vaart, A. and van der Laan, M. (2003). Smooth estimation of a monotone density, *Statistics* **37**, pp. 189–203.

van der Vaart, A. and Wellner, J. A. (1996). *Weak convergence and Empirical Processes* (Springer, New York).

Wahba, G. (1977). Optimal smoothing of density estimates, *Classification and clustering,* (ed.) J. Van Ryzin. Academic Press, New York , pp. 423–458.

Wahba, G. and Wold, S. (1975). A completely automatic french curve: Fitting spline functions by cross-validation, *Comm. Statist.* **4**, pp. 1–17.

Walker, A. M. (1971). On the estimation of a harmonic component in a time series with stationary independent residuals, *Biometrika* **58**, pp. 26–36.

Wand, M. P. and Jones, M. C. (1995). *Kernel Smoothing* (Chapman and Hall, CRC).

Watson, G. S. (1964). Smooth regression analysis, *Sankhyã* **A26**, pp. 359–372.

Watson, G. S. and Laedbetter, M. (1963). On the estimation of a probability density, *Ann. Math. Statist.* **34**, pp. 480–491.

Watson, G. S. and Laedbetter, M. R. (1964). Hazard analysis, *Biometrika* **51**, pp. 175–184.

Whittaker, E. T. (1923). On a new method of graduation, *Proc. Edinburgh Math. Soc. (2)* **41**, pp. 63–75.

Whittle, P. (1958). On the smoothing of probability density functions, *J. Roy. Statist. Soc., Ser. B* **20**, pp. 334–343.

Wold, S. (1975). Periodic splines for the spectral density estimation: The use of cross-validation for determining the degree of smoothness, *Comm. Statist.* **4**, pp. 125–141.

Index